China Internet Development Report 2021

Chinese Academy of Cyberspace Studies

China Internet Development Report 2021

Blue Book for World Internet Conference

中国工信出版集团 電子工業出版社 PUBLISHING HOUSE OF ELECTRONICS INDUSTRY http://www.phei.com.cn

Springer

Chinese Academy of Cyberspace Studies
Beijing, China

ISBN 978-981-19-9310-7 ISBN 978-981-19-9311-4 (eBook)
https://doi.org/10.1007/978-981-19-9311-4

Jointly published with Publishing House of Electronics Industry
The print edition is not for sale in China (Mainland). Customers from China (Mainland) please order the print book from: Publishing House of Electronics Industry.

This Springer imprint is published by the registered company Springer Nature Singapore Pte Ltd.
The registered company address is: 152 Beach Road, #21-01/04 Gateway East, Singapore 189721, Singapore

Preface

2021 saw the "Centenary of the Communist Party of China", the start of the "14th Five-Year Plan", and a crucial stage for building a moderately prosperous society in all respects and building a socialist modern country in all respects. Under the guidance of Xi Jinping's thoughts on socialism with Chinese characteristics in the new era, especially his important thought on building China into a strong cyberpower, the strategic deployment of "building China into a strong cyberpower" and the realization of the "Two Centenary Goals" are implemented simultaneously, making a series of new progress and achievement. We compile *China Internet Development Report 2021* (hereinafter referred to as "the *Report*") to showcase the achievements of China Internet development in 2021, envisage the future of China Internet development, and provide relevant parties with diverse materials and accurate data.

(1) Taking Xi Jinping's thoughts on socialism with Chinese characteristics in the new era, especially his important thought on building China into a strong cyberpower as fundamental guide. Since the 18th National Congress of the Communist Party of China, the CPC Central Committee with comrade Xi Jinping at the core has focused on the great struggle with many new historical characteristics, attached importance to the development and governance of the Internet, made a series of major decisions, and carried out a series of significant measures, in order to attain historic achievements in cyberspace administration in China and create a model of cyberspace governance with Chinese characteristics. In January 2021, *The Excerpts of Xi Jinping's Statements on Building China into a Strong Cyberpower* was officially published and distributed. It serves as an authoritative guide for us to learn General Secretary Xi Jinping's important thought on building China into a strong cyberpower, and possesses great significance for us to firmly promote the construction of a strong cyberpower. Taking General Secretary Xi Jinping's important thought on building China into a strong cyberpower as the cardinal principle, we comprehensively display the vivid example and the latest achievement in deeply learning and implementing General Secretary Xi Jinping's important thought on building China into a strong cyberpower.

(2) Deeply implementing the new philosophy of development and comprehensively representing the high-quality development of cyberspace administration.

The *Report* demonstrates the new development stage of cyberspace administration, and the latest practice of implementing the new philosophy of development, building the new pattern of development, and solidly promoting the "14th Five-Year Plan". In the past year, China Internet developed healthily, and Internet culture advanced vigorously, which ensured that positive energy and mainstream values prevailed. Digital economy played an increasingly important role in the dual-circulation of domestic and foreign markets. With new breakthroughs made in key and core technologies in the information field, the data element market developed steadily, the construction of digital society and digital government advanced continuously, and "information benefiting and serving people project" produced new results. The COVID-19 pandemic occurs frequently, the virus mutates quickly, and pandemic prevention and control paints a grim picture. Under such circumstances, digital technology robustly supports the control of infection, virus traceability, vaccine and drug R & D, and the resumption of work and production. Relevant departments continually strengthen the construction of cybersecurity guarantee system and capacity, make positive progress in rule-of-law construction in cyberspace, and effectively launch anti-monopoly on the platform economy and the prevention of the disorderly expansion of capital. In China Internet development, relevant parties take concerted efforts, communicate timely, and forge a coordinated system, which provides strong services, support, and guarantee for the comprehensive construction of a socialist modern country.

(3) Providing vivid examples in China Internet development in a comprehensive, objective, accurate, and adequate way. Focusing on China Internet development in 2021, the *Report* systematically evaluates Internet development in 31 provinces (autonomous regions and municipalities directly under the Central Government, excluding Hong Kong, Macao, and Taiwan) in China from six dimensions, i.e., infrastructure construction, innovation capacity, digital economy development, Internet application, cybersecurity and cyberspace governance, and reflects the Internet development level of various regions. This provides ideas for various regions to accurately utilize comparative advantages, further clarify the strategic objectives and priorities of Internet development, and boost the development of cyberspace administration toward the goal of building China into a strong cyberpower. The *Report* deeply analyzes new technologies, applications, and trends in information infrastructure construction, IT development, digital economy development, e-government development, Internet content construction, cybersecurity construction, rule-of-law construction in cyberspace and exchanges, and cooperation in international cyberspace governance.

We hope that the *Report* objectively records the progress of China Internet development; comprehensively presents new achievements, practices, and experiences of China Internet development; and provides references for readers to understand China Internet development.

Beijing, China
August 2021

Chinese Academy of Cyberspace Studies

Overview

2021 saw the "Centenary of the Communist Party of China", the start of the "14th Five-Year Plan", and a crucial stage for building a moderately prosperous society in all respects and building a socialist modern country in all respects. On the great occasion of the "Centenary of the Communist Party of China" and the historical convergence of "Two Centenary Goals", General Secretary Xi Jinping solemnly declares on behalf of the Party and Chinese people that: "after the continuous struggle of the whole Party and the people of all ethnic groups, we have realized the first centenary goal of building a moderately prosperous society in all respects. This means we have brought about a historic resolution to the problem of extreme poverty in China, and we are now marching with confidence towards the goal of building China into a great modern socialist country in all respects".

In the new development stage, under the guidance of Xi Jinping's thoughts on socialism with Chinese characteristics in the new era, especially his important thought on building China into a strong cyberpower, China implements new philosophy of development and forges new development landscape by upholding overall Party leadership, pursuing a people-centered approach to development, strengthening strategic coordination, promoting law-based cyberspace governance, and attaching equal importance to security and development. Additionally, China optimizes top-level design and strategic planning in cyberspace administration, and makes plans for Internet content construction and management, cybersecurity protection, IT-based development and international exchanges and cooperation in cyberspace, which establish a solid foundation for boosting the high-quality development of cyberspace administration and giving better play to the leading role of cybersecurity and informatization.

In line with *The Outline of the 14th Five-Year Plan for National Economic and Social Development and the Long-Range Objectives Through the Year 2035*, China will embrace the digital era; release the potential of big data; build China into a strong cyberpower; form new advantages in digital economy; accelerate the construction of digital society; improve the level of digital government; create healthy digital ecology; and reform the models of production, lifestyle, and governance with digital transformation. Presently, China continues to expand the scale of information infrastructure, raise IT innovation capacity, achieve remarkable results in the construction of digital society and digital government, quicken information-based services for the public, realize the innovation and development of Internet content, improve the cybersecurity guarantee capacity, and deepen international cooperation in the digital field. Therefore, digital technology becomes an important force to empower real economy, serve and benefit the public, make contributions to a decisive victory in poverty alleviation, and inject new momentum into economic development in China.

The international landscape proves increasingly complex, unstable, and ever-changing, and the COVID-19 pandemic has a far-reaching impact. The world economy plunges into the downturn and economic globalization runs up against headwinds. Amid the complex and changeable economic and political pattern in the world, unilateralism and protectionism pose a threat to global peace and development. China actively responds to new problems and challenges, unswervingly follows its development path, seizes the historical opportunities of an information revolution, enhances scientific and technological self-reliance and self-improvement, and forms a new pattern of cybersecurity. Besides, China vigorously promotes the Belt and Road Initiative, jointly builds a community with a shared future in cyberspace, and contributes Chinese wisdom and strength to world Internet development.

China Internet Development in 2021

In 2021, under the guidance of Xi Jinping's thoughts on socialism with Chinese characteristics in the new era, especially his important thought on building China into a strong cyberpower, China firmly grasped the historic opportunity to continuously promote the construction of information infrastructure, accelerated the innovation-driven R & D of information technology, rapidly developed digital economy, and improved information-based services for the public. Simultaneously, China strengthened Internet content construction, reinforced the law-based governance level and cybersecurity guarantee capacity in cyberspace, deepened international exchanges and cooperation in cyberspace, and attained a series of new achievements in Internet development.

With Overall Deployment of New-Type Infrastructure, the Construction of Information Infrastructure Steadily Advances

China accelerates the deployment of the broadband network, satellite Internet, and other information infrastructure; promotes the full coverage of optical-fiber broadband network in urban and rural areas; expedites the large-scale deployment of

IPv6 in infrastructure and industrial application; rapidly quickens the construction of computing power, new technologies, and application-oriented infrastructure; and facilitates the digital transformation of economy and society.

China makes positive progress in broadband network construction and builds the world's largest 5G Standalone (SA) and optical-fiber networks. Ministry of Industry and Information Technology issued *The Action Plan for Dual Gigabit Network Coordinated Development 2021–2023*, which proposed to accelerate the construction of a dual gigabit network. The total number of optical-fiber-to-home port reaches 899 million, covering 467 million users. Particularly, gigabit optical-fiber network covers more than 80 million users.[1] 5G mobile communication technology is extensively used in commercial cases for the first time. By June 2021, China has totally built 961,000 5G base stations, accounting for more than 70% of the world and fully covering all the cities above the prefecture level.[2] The large-scale deployment of IPv6 produces remarkable results, and the number of active user shoots up. By June 2021, the number of IPv6 address realized 62,023/32, and the number of IPv6 active user reached 533 million.[3] The data centers of three major telecom operators (i.e., China Mobile, China Telecom, and China Unicom) basically complete IPv6 transformation. In July 2021, Office of the Central Cyberspace Affairs Commission, National Development and Reform Commission, and Ministry of Industry and Information Technology jointly issued *The Notice on Accelerating and Promoting the Large-Scale Deployment and Application of Internet Protocol Version 6 (IPv6)* to comprehensively promote the large-scale deployment and application of IPv6 and accelerate Internet evolution and upgrading. China continues to strengthen policy guidance for new-type infrastructure construction and take the construction to the fast lane. According to the data of China Center for Information Industry Development (CCID), in 2020, China invested *c.* one trillion *yuan* to key fields of new-type infrastructure like industrial Internet, big data center, 5G, and AI. China formulates policies to foster the high-quality development of data center, launches "Channeling Computing Resources from the East to the West" Project, and accelerates the construction of national integrated big data center coordinated innovation system. In industrial Internet, the construction of networks, platforms, and security systems speeds up. By the end of March 2021, there were more than 100 platforms with industrial and regional influence, more than 70 million sets of connected equipment, more than 590,000 industrial apps, and more than one million cloudified enterprises, with the scale of industrial Internet industry reaching three trillion *yuan*.[4] National Development and Reform Commission and other ministries and commissions jointly issued *The Implementation Plan for Accelerating the Cultivation of New Forms of*

---

[1] Data Source: Ministry of Industry and Information Technology.

[2] The State Council Information Office: "The Press Conference on the Development of Industry and Information Technology in the First Half of the Year Held by the State Council Information Office". July 16, 2021, http://www.scio.gov.cn/xwfbh/xwbfbh/wqfbh/44687/46299/index.htm.

[3] Data Source: *The 48th Statistical Report on China Internet Development* released by China Internet Network Information Center.

[4] Data Source: Ministry of Industry and Information Technology.

Consumption, which planned to promote the deployment and application of IoV and charging pile (station). Since 2020, Ministry of Industry and Information Technology has approved Hunan Province (Changsha City) and Chongqing (Liangjiang New Area) to build national-level IoV pilot areas. Satellite Internet develops fast. In 2020, the output value of satellite navigation and location service industry totaled 403.3 billion *yuan*, up by 16.9% year on year.[5] In April 2021, China Satellite Network Group was established, which signaled a new stage in the development of satellite Internet.

Breakthroughs Are Constantly Made in Internet Information Technology and the Awareness and Capacity of Innovation Are Continuously Improved

China clings to promoting high-quality development via scientific and technological innovation; speeds up the advancement in basic theoretical research and basic algorithm development; steps up efforts in the development of key and core technologies; continues to augment the innovation capacity in new-generation information technology; and makes progress in the fields of high-performance computing, quantum communication, and operating system. Technological weaknesses are sustainably tackled, open-source ecosystem is further constructed, and a positive cycle of technological innovation and application is gradually formed.

Exascale supercomputing comes to a stage of technological breakthrough. In 2021, in *The 57th Global Top 500 Supercomputers List*, "Sunway TaihuLight" and "Tianhe-2A" of China ranked the 4th and the 7th, respectively. Memory industry develops rapidly and turns from technological breakthrough to capacity growth. OS product sees the stage of large-scale application, and Harmony OS, a full-scene distributed operating system, officially enters the stage of commercial promotion and application. Various countries progressively explore the cutting-edge technology of quantum information, and attain prominent results in quantum communication, computing, and measurement. Open-source ecology improves and open-source product diversifies, with Chinese Internet enterprises successively shortlisted in global major open-source contributors. New technologies and applications continuously emerge. Applications like automatic driving and vaccine and drug R & D quickly turn to reality and meet the urgent needs of society.

---

[5] Data Source: *2021 White Paper on the Development of China's Navigation Satellite and Location Service Industry* published by GNSS and LBS Association of China.

Digital Economy Plays a Leading Role in Economic Development and New "Dual Circulation" Development Pattern Accelerates

China vigorously bolsters the innovation and application of new technologies like 5G, IoT, cloud computing, big data, AI, and blockchain. Various regions strengthen the deployment of the digital economy, stimulate the development of the digital economy, and transform or upgrade traditional industries. As digital transformation quickens, digital technologies deeply integrate with real economy, and the industrial digitalization and digital industrialization score a great success, which inject new momentum to the stable development of national economy.

Over the past year, under the dual impact of the COVID-19 pandemic and global economic downturn, the digital economy in China thrived against headwinds. As China Academy of Information and Communications Technology (CAICT) estimated, in 2020, the scale of the digital economy in China reached 39.2 trillion *yuan*, accounting for 38.6% of the total GDP. With a high growth rate of 9.7%, digital economy becomes a key force that drives stable economic growth. Structurally, the scale of digital industrialization realizes 7.5 trillion *yuan*, which continually breeds new industries as well as new business forms and models, and connects to a global high-end industrial chain. The scale of industrial digitalization continues to expand, reaching 31.7 trillion *yuan*. The digital level of agriculture, industry, and service sectors keeps rising.[6] E-commerce develops stably, online service and e-commerce industries see explosive growth, and the scale of social e-commerce continuously enlarges. In 2020, the volume of e-commerce transaction in China achieved 37.21 trillion *yuan*, with a year-on-year increase of 4.5%. The business revenue of e-commerce service industry totaled 5.45 trillion *yuan*, with a year-on-year increase of 21.9%.[7] China steps up efforts in the construction of free trade zone. In 2020, China issued *The Master Plan for the Construction of Hainan Free Trade Port*, which planned to construct three free trade zones in Beijing, Hunan, and Anhui and highlight the digital economy and cross-border e-commerce. In China, cross-border e-commerce maintains a high growth rate, and high-end manufacturing export displays a prominent trend. According to the statistics released by the General Administration of Customs, in the first quarter of 2021, China's cross-border e-commerce imports and exports realized 419.5 billion *yuan*, with a year-on-year increase of 46.5%. In type, products developed from low-price, high-frequency consumer goods into high-end manufacturing goods. The pilot of e-CNY advances in an all-round way, and pilot cities and areas continuously expand. The pilot basically covers the Yangtze River Delta, the Pearl River Delta, Beijing-Tianjin-Hebei as well as central, western, northeastern, and northwestern regions. By June 2021, more than 1.32 million pilot use cases were created, more than 20.87 million personal wallets were opened, and

---

[6] Data Source: *Digital China Development Report 2020* published by Cyberspace Administration of China.

[7] Data Source: *China E-Commerce Report 2020* published by Department of E-Commerce and Information Technology of Ministry of Commerce.

the accumulated transaction amount was *c.* 34.5 billion *yuan.*[8] Sharing economy robustly shores up the adjustment of economic structure. In 2020, the transaction volume in sharing economy market approximated to 3,377.3 billion *yuan.* The market scale of shared knowledge and skill, medical care, and other fields substantially grew, while that of shared accommodation, office, and travel reduced sharply over 2020.[9] The thriving digital economy fuels the rapid growth of new market entities; provides a large number of employment posts; and becomes an important channel for ensuring security in job, basic living needs, and operations of market entities. In 2020, Ministry of Human Resources and Social Security, together with relevant departments, released the list of 25 new occupations, e.g., intelligent manufacturing engineering and technology personnel, industrial Internet engineering and technology personnel, and online delivery personnel.[10] China invests more efforts to the training of personnel in digital economy. In January 2021, Ministry of Education and Academic Degrees Committee of the State Council approved the establishment of interdisciplinary subjects. In particular, the first-level disciplines like "Integrated Circuit Science and Engineering" were set up. This helps to construct an innovative-talent-training system that facilitates the rapid development of integrated circuit industry and cultivates urgently needed innovative talents for industrial development.

Information-Based Services for the Public Release New Vitality and Online Services Reach the Leading Level in the World

Under the impact of the COVID-19 pandemic, China implements the people-centered philosophy of development, accelerates the integration of digital technologies with socializing network and daily life in an all-round way, constantly improves "Internet + public services", promotes the construction of smart city and digital village, strengthens the opening and sharing of public data, optimizes the construction of digital government, and meets people's needs for a better life in a sustainable way.

By June 2021, the number of Internet user in China had hit 1.011 billion, with the Internet penetration rate reaching 71.6%.[11] In order to adapt to new trend in which digital technologies fully integrate with socializing network and daily life,

---

[8] Data Source: *The Research and Development Progress of China's e-CNY* published by the People's Bank of China.

[9] Data Source: *China's Sharing Economy Development Report 2021* published by State Information Center.

[10] Data Source: *Digital China Development Report 2020* published by Cyberspace Administration of China.

[11] Data Source: *The 48th Statistical Report on China Internet Development* released by China Internet Network Information Center.

China actively develop public services such as online classroom, Internet hospital, and smart library. 325 million and 239 million Internet users in China had access to online education and online medical service, respectively, accounting for 32.1% and 23.7% of all Internet users, respectively.[12] Internet hospital sees a rapid growth. Nearly 90% of physical medical institutions and Internet enterprises actively deploy and develop Internet hospitals. By the end of 2020, the number of Internet hospital in China reached 1,004, and the annual number of people diagnosed and treated increased by 150.0% year on year.[13] Smart city develops in an all-round way and extends to counties and communities. Various countries launch the pilot construction of smart city, smart community, future community, and smart property management. To comprehensively deploy the construction of digital village, Office of the Central Cyberspace Affairs Commission and other six departments have jointly issued *The Notice on Implementing the Pilot Project of National Digital Village* and chosen 117 counties (cities and districts) as the first-batch national digital village pilot areas. The construction of rural information infrastructure continuously improves, and more than 98% of administrative villages get access to optical-fiber broadband and 4G networks. By 2020, the online sales of rural products amounted to 1.79 trillion *yuan*, and 454,000 rural information-based service stations for farmers came into operation.[14] The construction of agricultural and rural big data achieves initial success. National Agricultural Products Quality and Safety Traceability Management Platform is widely promoted and applied, and the construction of planting, animal husbandry, and fishery information systems is continuously ameliorated. The construction of digital government in China solidly progresses, and the level of Internet-based government services further improves. "One-website handling", "handling affairs in various cities", and "handling affairs in various provinces" form a new trend and actualize "government services via mobile phones or apps". According to *2020 The UN E-Government Survey* released by the United Nations, China's global ranking in E-government Development Index jumped from the 65th in 2018 to the 45th in 2020. In particular, in Online Service Index, a core index that measures the development of national E-government, China leaped to the 9th place in the world. National Integrated Government Services Platform basically takes shape and plays a prominent role, which covers more than 5.67 million matters handled by 45 departments directly under the State Council, 31 provinces (autonomous regions and municipalities directly under the Central Government) and Xinjiang Production and Construction Corps, and around 1.16 billion government-service affairs coordinated by various departments.[15]. In the COVID-19 pandemic, relevant departments introduced "health codes" in the pandemic prevention and control, which took effect

---

[12] Data Source: *The 48th Statistical Report on China Internet Development* released by China Internet Network Information Center.

[13] Data Source: *2021 China Internet Hospital Development Report* published by National Telemedicine and Connected Healthcare Center and CN-Healthcare.

[14] Data Source: *Digital China Development Report 2020* published by Cyberspace Administration of China.

[15] Data Source: Website of National Integrated Online Government Services Platform, August 30, 2021, http://gjzwfw.www.gov.cn/col/col375/index.html

more than 40 billion times and contributed to the realization of "one code pass" with health code across the country.

The Construction of Internet Content Constantly Improves and Content Production Model Realizes Innovation and Development

General Secretary Xi Jinping reiterates that relevant departments should strengthen Internet content construction; propagate Internet positive energy; encourage the innovation on the idea, content, form, and approach in terms of Internet dissemination; highlight the timing, extent, and efficiency of Internet content; and consolidate the common ideological foundation upon which all people are united and work together. China accelerates and deepens media convergence, and establishes omni-media communication system. Internet users voluntarily participate in the construction of positive content, and take various means of communication to enhance the capacity of international communication. High-quality Internet cultural products embody the spirit of "thinking outside the box" and mark new development stage of the construction of Internet content.

Over the past year, the Internet saw the mass campaign for learning, propagating, and interpreting Xi Jinping's thoughts on socialism with Chinese characteristics in the new era, especially his important thought on building China into a strong cyberpower. In line with *The Excerpts of Xi Jinping's Statements on Building China into a Strong Cyberpower*, news websites and theory-centered websites fully utilized their strengths as well as the means of new media communication, and launched a series of new-type omni-media products that appealed to Internet users, youngsters in particular. This forged a strong learning atmosphere among the public. Relevant parties gave full play to the role of mainstream media platforms, deeply disseminated the great achievements of the Communist Party of China in 100 years after its founding as well as the outstanding deeds of advanced Party members, and focused on the plan and publicity of major events like "China's complete victory in the fight against poverty" and "the start of the 14th Five-Year Plan", which enlarged the influence of mainstream public opinion. Simultaneously, relevant departments furthered the "China Positive Energy Outstanding Works Collection, Selection and Exhibition Project", which staged many outstanding figures and works that played a guiding role in disseminating positive energy on the Internet in the year. As an effective means, short videos made sustainable progress in spreading positive energy. By June 2021, the scale of short video user in China reached 888 million, accounting for 87.8% of all Internet users.[16] Short video platforms, such as TikTok, Kuaishou, and bilibili, demonstrated their prowess in creating and spreading short videos under the theme

[16] Data Source: *The 48th Statistical Report on China Internet Development* released by China Internet Network Information Center.

of celebrating the Centenary of the Communist Party of China. Relevant departments guided platform users to create a substantial number of outstanding micro-documentaries and short videos and played an important part in online publicity of the CPC history learning and education. Internet content products present a user-centered trend, and online literature and online games maintain rapid growth. Media convergence deepens. County-level media convergence achieves the turn from massive surge to quality and efficiency-oriented development. Various regions establish radio and TV media-convergence development and innovation centers. Internet media technology facilitates the iteration and renewal of Internet content products, cultural industry keeps thriving, and the use case of "5G +UHD" technology diversifies. In December 2020, Migu Music Awards completed the global live broadcast of "5G + 8K + cloud performance" for the first time. In 2021, CCTV Spring Festival Gala adopted 8K UHD video live broadcast for the first time and combined "AI + VR" naked-eye 3D studio technology. China continues to launch Chinese Culture New Media Communication Project. *Luo Shen Shui Fu*, a dance program, fully integrates high-tech with traditional Chinese culture. According to official statistics, within 10 days of the broadcast, related topics on microblogs were read around five billion times, with 19 times of hot topics in 48 hours and more than 100 million viewers. This successfully disseminated the essence of excellent Chinese traditional culture. Additionally, central news websites centered on major events, focused on bilingual short videos, and carried out "My Love for the Communist Party of China" Global Internet Promotion Activity. This enhanced the interaction of cultural communication products, spread China's development stories, and demonstrated China's new changes in an all-round way.

With Diverse Threats to Cybersecurity, the Construction of Data Security Governance System Accelerates

Confronted with severe and complex cybersecurity landscape at home and abroad, China upholds a holistic approach to national security, continuously improves cybersecurity-related top-level design and system-standard construction of major fields like data security, strengthens the R & D of key cybersecurity technologies, solidifies the comprehensive competitiveness of cybersecurity industry, enhances cybersecurity publicity and education and personnel training, and achieves remarkable results in cybersecurity governance.

Noticeably, with the continuous development of new technologies and applications, cybersecurity faces new threats, risks, and challenges. In the first half of 2021, *c.* 23.07 million malicious program samples were captured, which spread more than 5.82 million times in a day on average. Around 4.46 million hosts were infected with malicious programs in China, with a year-on-year increase of 46.8%. The risks in cybersecurity vulnerabilities remain large. In the first half of 2021,

China National Vulnerability Database (CNVD) newly recorded 13,083 general-purpose security vulnerabilities, including 7,107 "zero-day" vulnerabilities, with a year-on-year growth of 55.1%. The large-scale development of new technologies and applications, e.g., cloud computing and industrial Internet, triggers new cyber-security problems. The number of cybersecurity incident on cloud platform remains large, and the proportion of industrial control equipment and system directly exposed to the Internet proves high. The leakage of user's personal information and data becomes prominent, and security risks on WeChat applets and IoV arouse great concern. China makes solid progress in building cybersecurity guarantee system. The Standing Committee of the National People's Congress deliberated and passed *The Data Security Law of the People's Republic of China*, the basic law in the data field in China. The State Council issued *The Regulations on the Security Protection of Critical Information Infrastructure,* which provided strong legal guarantee for deeply implementing the security protection of critical information infrastructure. Simultaneously, Cyberspace Administration of China and other ministries and commissions jointly released *Several Regulations on the Management of Automobile Data Security (For Trial Implementation), The Regulations on the Management of Vulnerabilities of Cybersecurity Products*, and other documents. Relevant departments tighten up cybersecurity review on platform enterprises such as "Didi Chuxing, "Yunmanman", "Huochebang", and "BOSS Zhipin". Cybersecurity market comes to a stage of rapid growth. With the ecological environment of cybersecurity industry continuously optimized, the construction of cybersecurity industrial park achieves initial results. Relevant departments deepen the training of cybersecurity talents and launch "Cybersecurity 10,000 Talents Training Funding Program" and other projects to tackle the lack of cybersecurity talents. National Cybersecurity Publicity Week is organized in various regions to intensify the awareness of cybersecurity among the public.

The Legalization in Cyberspace Comprehensively Advances and the Governance of Platform Economy Reaches a New Level

The CPC Central Committee issues *The Implementation Outline for the Construction of Law-Based Society 2021–2025*, which comprehensively deploys "law-based cyberspace governance". As stated, China will establish and improve comprehensive cyberspace governance system, and strengthen the law-based governance of cyberspace, law-based construction of website, and law-based use of Internet, in a bid to promote the legalization of cyberspace in an all-round way and create a clear cyberspace. China continues to optimize legal system for cyberspace governance; turns from regulation and principle-oriented legislation to niche and vertical filed-oriented legislation; highlights the law enforcement, supervision, and inspection in cyberspace; and takes various forms to cultivate citizen's awareness of law in cyberspace.

The legalization in cyberspace continually advances, and the legislative process in key fields like personal information protection accelerates. In August 2021, the Standing Committee of the National People's Congress voted to adopt *The Personal Information Protection Law of the People's Republic of China*, which specifically regulated social concerns such as excessive collection of personal information by apps and automated decision-making by using personal information. Meanwhile, the Supreme People's Court released *The Provisions of the Supreme People's Court on Several Issues Concerning the Application of Law in the Trial of Civil Cases in Relation to the Use of Face Recognition Technology to Process Personal Information.* In order to effectively regulate the collection of personal data on apps, Cyberspace Administration of China and other ministries and commissions jointly issue *The Regulations on the Scope of Necessary Personal Information for Common Types of Mobile Internet Applications*; launch the rectification of illegal collection and use of personal information on apps; conduct special inspection and evaluation on more than 1,000 apps of 14 common types such as input method and map navigation that are downloaded by users in large quantities; and announce 351 apps with serious violations of laws and regulations, ordering them to rectify within the time limit. Apps that are not rectified within the specified time limit will be removed, which achieve satisfactory results.[17] Relevant departments strengthen the regulation of platform economy and crack down on monopoly and unfair competition. To this end, they publish *The Antitrust Guidelines on Platform Economy*, *The Measures for the Supervision and Administration of Online Transactions* and other documents, and impose administrative penalties on Alibaba Group for its "one out of two" monopoly according to law. Besides, they issue *The Guiding Opinions on Strengthening the Standardization and Management of Online Livestreaming* to regulate online livestreaming behaviors. Relevant departments also consolidate the protection of the minors in cyberspace. For instance, *The Minors Protection Law of the People's Republic of China* adds a chapter of "Internet Protection", and Beijing Internet Court establishes the first Internet Juvenile Court in China, which demonstrates China's determination to protect the rights and interests of the minors in cyberspace. Relevant departments fortify the law enforcement in cyberspace, tackle various malpractices in cyberspace, initiate "Internet clean-up campaign", deepen the centralized rectification of black industries like camera peeping, vigorously combat the chaos of undesirable fan culture and mobile app information content, and improve comprehensive cyberspace governance capacity. According to the statistics, in 2020, national cyberspace administration organs summoned 4,282 websites for talks according to law, gave warnings to 4,551 websites, and closed 158,000 account groups in violation of laws and regulations according to user-service agreements.[18] China actively carries out law popularization in cyberspace, which works

[17] Data Source: Cyberspace Administration of China.

[18] Data Source: Cyberspace Administration of China.

in the legislation, law enforcement, and judicature. The idea that "cyberspace is not a place beyond the rule of law" is widely recognized. China formulates the "8th Five-Year Plan" for law popularization; strengthens the functions of national smart law-popularization platforms; establishes omni-media law- communication system; gives full play to the role of governments, enterprises, and social organizations; and enhances the awareness of the law among the public. Additionally, China boosts the construction of integrity in cyberspace, stages Internet Integrity Conference, advocates Internet civilization, creates a good atmosphere for everyone to offer suggestions and participate in Internet-integrity construction, and jointly forges a cyberspace environment featuring integrity and trustworthiness. Relevant departments step up efforts to receive and handle the reporting of illegal and harmful information. In 2020, departments in charge of Internet reporting at various levels received 163 million reports, with a year-on-year increase of 17.4%.[19] In China, the number of participant in Internet rumor governance continuously expands. The joint rumor refutation mechanism led by China Internet Joint Rumor Refutation Platform and co-launched by central media platforms and provincial rumor refutation platforms plays a prominent role in Internet rumor governance.

International Exchanges and Cooperation in Cyberspace Deepen and the Proposition on Data Security Becomes a Global Consensus

General Secretary Xi Jinping stresses that: "China stands ready to work with international community to seize the historical opportunity of information revolution, cultivate new momentum for innovation and development, create new landscape of digital cooperation, establish new pattern of cybersecurity, build a community with a shared future in cyberspace, and join hands to create a better future for mankind". Over the past year, China continued to deepen the process of international exchanges and cooperation in cyberspace; promote the formulation of international rules on digital development and cyberspace with the United Nations as main channel and *Charter of the United Nations* as basic principle; provide developing countries with digital assistance like technology, equipment, and services to enable them to share the dividends of the digital era, and foster the implementation of *Transforming our World: The 2030 Agenda for Sustainable Development* of the United Nations.

Under the framework of the United Nations, China plays an active role in making international rules in cyberspace. China puts forward *Global Initiative on Data Security*. China's proposal of establishing supply-chain security-risk evaluation mechanism is acknowledged by the U. N. Information Security Group of Governmental Experts (GGE) in the final report. On the basis of such platforms as the United Nations Internet Governance Forum (IGF), World Summit on the Information Society (WSIS), and World Economic Forum (WEF), China deeply partakes in multilateral

[19] Data Source: WeChat official account of Cyberspace Administration of China.

activities of international cyberspace governance, hosts World Internet Conference for consecutive years, builds cooperation platforms, and discusses the development trend of cyberspace with international community. China strengthens the interpretation of ideas on cyberspace, releases the action initiative *Jointly Build a Community with a Shared Future in Cyberspace*, and compiles relevant textbooks on the cases of cyberspace governance, with the initiative widely recognized by international community. China facilitates implementation of *Transforming our World: The 2030 Agenda for Sustainable Development* of the United Nations, and promotes eradicating poverty with digital assistance. At the United Nations Human Rights Council, APEC Digital Poverty Alleviation Seminar, and other meetings, China shares the experience in digital poverty alleviation and the models of minors Internet protection, prepares to establish an international research center for big data-based sustainable development, and boasts the first country in the world that serves the UN sustainable development goal with big data. China cements and bolsters the exchanges and cooperation in cyberspace, and promotes multilateral cooperation in cyberspace among G20, APEC, BRICS, Shanghai Cooperation Organization, and ASEAN. Simultaneously, China releases a series of cooperation declarations, such as *The Declaration of the Ministers of G20 on Digital Economy*, *China-ASEAN Initiative on Building Digital Economy Partnership*, and *Initiative on China-Africa Jointly Building a Community with a Shared Future in Cyberspace*. China formally signs *Regional Comprehensive Economic Partnership* (RCEP) with relevant countries. Various parties reach consensus on e-commerce rules, cross-border data flow, and other matters, which effectively accelerates the development and cooperation of regional digital economy. China strengthens bilateral exchanges and cooperation with the United States, Russia, Europe, and other countries; deepens exchanges and cooperation with neighboring countries on cyberspace administration; and advances the cooperation with BRI countries and regions in digital economy and IT-based construction. China and the United States launch high-level strategic dialogues to enhance mutual understanding. China and Russia extend *China-Russia Treaty of Good-Neighborliness and Friendly Cooperation* and expand the cooperation in digital economy and other fields. China and League of Arab States jointly release *China-League of Arab States Cooperation Initiative on Data Security*, and League of Arab States becomes the first regional organization in the world to jointly publish the initiative on data security with China. China and African Union sign *The Plan for the Cooperation between the Government of the People's Republic of China and African Union on Jointly Promoting the Construction of the Belt and Road Initiative*. The two sides reinforce the cooperation in the fields of information infrastructure construction, industrial digitalization, and information technology services, and work together to seize the opportunities for the development of digital economy.

Information Technology Plays a Leading Role in the Reform of Public Welfare Model and the Form of Internet Public Welfare Diversifies

The Outline of the 14th Five-Year Plan for National Economic and Social Development states that China will boost the development of charity and regulate the development of Internet charity platforms, highlighting the importance of developing Internet public welfare. *Vis-à-vis* traditional forms of public welfare, Internet public welfare improves the transparency, convenience, and accessibility of participation. Public welfare activities feature diverse forms and scenes, and various public welfare products mushroom, with innovation on public welfare projects and more channels for public participation.

Characterized with decentralization, openness and transparency, information traceability, and automatic execution via smart contract, blockchain technology effectively improves the information transparency of public welfare projects and helps to address the problem of public trust. In December 2020, China Internet Development Foundation hosted "2020 Blockchain + Charity Seminar" and jointly discussed the use of blockchain technology to innovate a new model of "link-based public welfare", via which new technologies can empower public welfare and enhance public trust.[20] By dint of platform advantages and traffic resources, Internet enterprises continue to expand the scope of public welfare. In particular, public welfare-oriented consumption emerges and becomes a way for the public to participate in Internet public welfare. Tencent Gongyi launches "99 Giving Day", in which the number of participant and the scale of donation continue to rise. In 2020, 57.8 million people took part in "99 Giving Day", who donated 2.32 billion *yuan*. With 324 million *yuan* from enterprises and 400 million *yuan* from Tencent Charitable Foundation, a total of 3.044 billion *yuan* was raised. In July 2021, the heavy rain caused floods in Henan. In order to help the disaster areas in Henan, Sina Weibo Public Welfare Platform quickly launched a number of flood-control public welfare projects. By July 22, a total of 25 projects were started, in which 1,446,400 people totally donated 29.8472 million *yuan*.[21] The proportion of small-amount donor increases significantly, and public welfare projects receive more attention and resources support. More innovative and efficient public welfare projects come to rise, which diversifies the ecology of Internet public welfare.

[20] Cyberspace Administration of China: "Productive Results: Illuminating Views on 'Blockchain + Charity Seminar'". December 31, 2020, http://www.cac.gov.cn/2020-12/31/c_1610983882205930.htm.

[21] Data Source: Sina Weibo Public Welfare Platform, July 22, 2021.

Internet Development in Various Provinces (Autonomous Regions and Municipalities Directly under the Central Government)

Since 2017, *China Internet Development Report* has released China Internet Development Index. Guided by General Secretary Xi Jinping's important thought on building China into a strong cyberpower, China Internet Development Index aims to conduct comprehensive evaluation of the achievements and levels of Internet development in various provinces (autonomous regions and municipalities directly under the Central Government, excluding Hong Kong, Macao, and Taiwan), by building an objective, truthful, and accurate comprehensive evaluation index system. In this way, various regions can further clarify the strategic objectives and priorities of Internet development, accurately utilize comparative and regional advantages in Internet development, and promote building China into a strong cyberpower in cyberspace administration.

China Internet Development Index System covers six aspects, i.e., information infrastructure construction, innovation capacity, digital economy development, Internet application, cybersecurity, and cyberspace governance. It comprehensively showcases Internet development of 31 provinces (autonomous regions and municipalities directly under the Central Government) in China, and provides a quantifiable reference for them in Internet development. Grounded in the index system in 2020, the evaluation index in 2021 takes into consideration Internet development in various regions; widely solicits opinions from national departments, provinces, and cities; think tanks on cyberspace administration and experts in relevant fields; and redesigns or improves the index system in 2020 (see "Overview" Table 1).

In order to ensure the authenticity, entirety, and accuracy of the data, the evaluation data of China Internet Development Index in 2021 mainly stem from the statistical data of Ministry of Science and Technology, Ministry of Industry and Information Technology, Ministry of Commerce, General Administration of Customs, Cyberspace Administration of China, National Bureau of Statistics and other departments and institutions, as well as the statistical data of relevant cyberspace administration departments in various provinces (autonomous regions and municipalities directly under the Central Government).

Comprehensive Rankings of China Internet Development Index in 2021

In line with China Internet Development Index System in 2021, the list of top ten regions in Internet Development Index of 31 provinces (autonomous regions and municipalities directly under the Central Government) in China is shown in "Overview" (Fig. 1).

Table 1 China Internet Development Index System

First-Level Index	Weight	Second-Level Index	Description
Infrastructure Construction Index	0.18	Broadband Infrastructure	Per capita Internet broadband access port, the proportion of optical-fiber user, fixed broadband network rate, etc.
		Mobile Infrastructure	The download rate of 4G network, the proportion of 4G mobile phone user, the number of 5G base, etc.
		Application Infrastructure	The proportion of IPv6 active terminal, the proportion of IPv6 traffic, the number of industrial Internet platform, the competitiveness of new-type infrastructure
Innovation Capacity Index	0.18	The Environment for Innovation	The total number of incubated enterprise, the number of R & D personnel, big data talent reserve, etc.
		Innovation Input	The proportion of R & D expenditure to GDP, the proportion of corporate R & D personnel, the proportion of governmental R & D investment to GDP, etc.
		Innovation Output	The number of scientific research paper per 10,000 people, the number of national award for scientific & technological achievement per 10,000 people, the number of invention patent per 10,000 people, etc.
Digital Economy Development Index	0.19	Basic Index	The number of Internet user per 10,000 people, the total business of telecommunications service, the evaluated value of added value of information transmission, software and IT services, etc.

(continued)

Table 1 (continued)

First-Level Index	Weight	Second-Level Index	Description
		Integrated Index	The readiness rate of intelligent manufacturing, the numerical control rate of key processes, the digitization rate of production equipment, the proportion of e-commerce consumption expenditure to the final consumption expenditure, etc.
		Industrial Index	Big data industry development index, the competitiveness of AI-technology industrial regions, the number of enterprise listed in *China Top 100 Internet Enterprises*, the scale of e-commerce transaction, the number of unicorn, etc.
Internet Application Index	0.18	Personal-Level Application	The traffic of mobile traffic per household, the penetration rate of mobile phone, etc.
		Business-Level Application	The application rate of industrial cloud platform, the proportion of networked collaborative enterprise, the proportion of service-oriented manufacturing enterprise, the proportion of enterprise engaged in e-commerce transaction, etc.
		Public Application	The e-government-service capacity of provincial government, the digital level of pandemic prevention and control, etc.
Cybersecurity Index	0.15	The Environment for Cybersecurity	The number of controlled malicious computer program, the number of implanted backdoor website and tampered website, the number of handled cybersecurity incident, etc.

(continued)

Table 1 (continued)

First-Level Index	Weight	Second-Level Index	Description
		The Construction of Cybersecurity Awareness	The number of participant in a series of Cybersecurity Week Activities, the construction of cybersecurity personnel, cybersecurity search index, cybersecurity information index, page views of cybersecurity we-media, etc.
		The Development of Cybersecurity Industry	The number of cybersecurity enterprise, the number of enterprise listed in *Top 100 Cybersecurity Enterprises*, etc.
Cyberspace Governance Index	0.12	Social Collaboration	The number of Internet-related social organization, the number of authorized size at provincial cyberspace administration department, etc.
		Content Dissemination	The communication effect of provincial government-service microblog, the communication effect of provincial government-service WeChat account, etc.
		Law-Based Cyberspace Governance	Cyberspace-related strategies, policies, regulations and normative documents, the number of Internet news information license, the evaluation of cyberspace governance level, etc.

In general, various provinces (autonomous regions and municipalities directly under the Central Government) comprehensively and thoroughly implement General Secretary Xi Jinping's important thought on building China into a strong cyberpower, earnestly act in accordance with the deployment requirements of the CPC Central Committee, give full play to their own advantages, and vigorously expedite the development process of Internet based on regional advantages or local conditions. 2021 signaled the start of the second centenary goal and the "14th Five-Year Plan". Tense moments prevail in global political relations, and the COVID-19 pandemic

Table 2 Top Ten Provinces (Autonomous Regions and Municipalities Directly under the Central Government) in the Second-Level Evaluation Indexes in 2021

Ranking	Infrastructure Construction Index	Innovation Capacity Index	Digital Economy Index	Internet Application Index	Cybersecurity Index	Cyberspace Governance Index
1	Guangdong	Guangdong	Beijing	Beijing	Beijing	Hebei
2	Beijing	Beijing	Zhejiang	Shanghai	Guangdong	Shanghai
3	Zhejiang	Jiangsu	Shanghai	Shandong	Shanghai	Beijing
4	Shandong	Zhejiang	Jiangsu	Guangdong	Shandong	Sichuan
5	Shanghai	Shanghai	Guangdong	Sichuan	Sichuan	Tianjin
6	Henan	Shandong	Shandong	Chongqing	Fujian	Guangdong
7	Tianjin	Hubei	Sichuan	Fujian	Jiangsu	Hubei
8	Jiangsu	Anhui	Henan	Anhui	Hubei	Shandong
9	Liaoning	Tianjin	Fujian	Zhejiang	Chongqing	Henan
10	Fujian	Fujian	Chongqing	Tianjin	Anhui	Sha'anxi

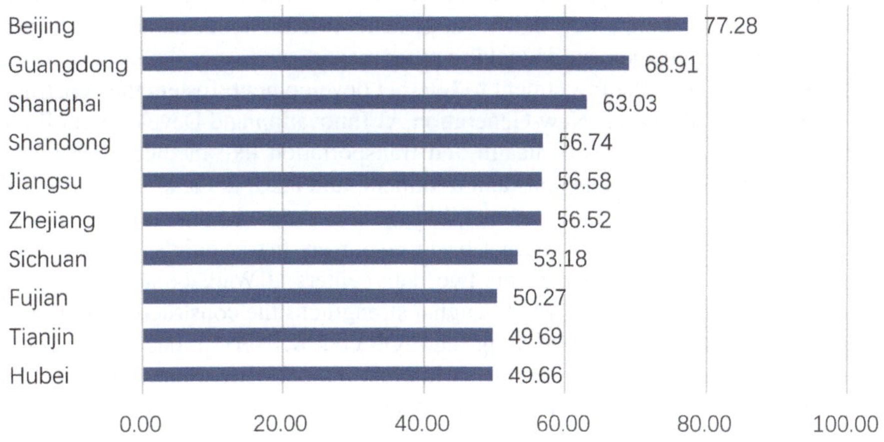

Fig. 1 The List of Top Ten Regions in China Internet Development Index

has a far-reaching impact on citizen's life. Against such backdrop, local governments surmount numerous difficulties, resolutely implement the guiding spirit of the CPC Central Committee, further strengthen Internet content construction and management, boost the development of digital economy and digital society, and reinforce cybersecurity guarantee capacity. In Comprehensive Rankings of China Internet Development Index in 2021, top ten provinces (municipalities directly under the Central Government) in Internet development are Beijing, Guangdong, Shanghai, Shandong, Jiangsu, Zhejiang, Sichuan, Fujian, Tianjin, and Hubei.

Based on the functional orientation as capital, Beijing actively adapts to new demand for high-quality development and takes "people-oriented and data-empowered" development as core idea. By improving urban governance and service level and promoting the deep integration of new-generation IT with urban governance and public services, Beijing continuously strengthens digital governance, greatly advances the modernization of urban governance system and governance capacity, actively enhances the construction of national technological innovation center, comprehensively deploys future intelligent system platform, and timely accelerates the construction of digital trade pilot area.

Guangdong steadily promotes the construction of basic telecommunications network and accelerates the deployment of application-oriented infrastructure, ranking the 1st in the number of national industrial Internet platform in China. Guangdong scientifically plans the construction of data center; highlights the R & D of basic software and hardware at cutting-edge industrial chains; and quickens the digital, networked, and intelligent development of Internet. Besides, Guangdong furthers the legislation in the field of digital economy, formulates *The Interim Measures for the Opening of Public Data in Guangdong Province*, optimizes the construction of data-trading venues and supporting institutions, and releases the value of public data resources by establishing public data management evaluation mechanism, exploring public-data-asset management and taking other means.

Shanghai increases the investment to Internet development, fosters the construction of National (Shanghai) New-Generation AI Innovation and Development Pilot Area, deploys the use cases of health and transportation like medical data interconnection and auxiliary diagnosis and treatment, and takes the lead in developing AI governance research in the pilot area. Shanghai expands the construction, popularization, and application of Internet basic resources, firstly builds "dual gigabit broadband city" in China, and forms two data centers at Waigaoqiao Free Trade Zone and Baosteel Industrial Park. Shanghai strengthens the construction of digital society, creates the first "QR code suing" and "QR code delivery" models in the court system, and gives full play to the role of Internet in social life. Shanghai launches the pilot work of e-CNY, deepens inclusive finance, and extends the service models of digital finance.

Shandong speeds up the development of industrial Internet, vies for the leading position in the construction of industrial Internet, takes the lead in establishing the incentive mechanism of "modern competitive industrial cluster + artificial intelligence" in China, and releases policies on "Internet + advanced manufacturing industry" and documents like *Several Measures to Accelerate the Development of Industrial Internet*. Shandong comprehensively promotes the construction of new-type smart city, builds new-type smart city ecosystem that features "five perceptive abilities, four basic capabilities and convenient services", constructs smart communities (villages), and completes "the last mile" of smart and convenient services. Shandong implements intelligent and green technological transformation project and industrial Internet "One Hundred" Project, accelerates the digital transformation of manufacturing enterprises in an all-round way, launches "Ai Shan Dong" app (a general portal of mobile government services), and deepens "Internet + government services".

Jiangsu constructs 5G industrial structure map online system; integrates IoT perception facilities and communications systems into the unified planning and construction of public infrastructure; applies 5G into IoT, IoV, industrial Internet, and smart city; and quickens the construction of 5G industrial ecology. Simultaneously, Jiangsu promotes the implementation of big data development action plan; formulates big data industrial map; steps up efforts in the construction of new-type infrastructure in the fields of 5G network, digital transportation, smart health, and IoV; plans and implements 20 key projects in "5G + industrial Internet"; starts the high-level construction of provincial IoV pilot areas in Wuxi, Suzhou, Nanjing, and Nantong; and expedites the applications of technological products like car chip, intelligent perception, edge computing, and vehicle-road collaboration.

Zhejiang makes full use of digital technologies like cloud computing, IoT, big data, mobile Internet and AI in the fight against the pandemic, vigorously popularizes online education, Internet medical care, online entertainment and other forms of people's livelihood services, builds modern government of "intelligence-based governance and practical-spirit-based innovation", and comprehensively boosts the digital transformation of government. By establishing the county-level "one database, one map, one code and one index" system with the coordination among five-color "pandemic map", "health code" and "precision and intelligent control index", Zhejiang actualizes systematic and accurate digital governance, which facilitates citizen's life and enterprise's resumption of work, and improves the efficiency of management and control to the maximum extent.

Sichuan enhances the construction of rural communications infrastructure, overcomes the disadvantages in rural network, and achieves the full coverage of optical-fiber access and 4G network in all administrative villages. Sichuan vigorously promotes the construction of 5G network; realizes the full coverage of 5G network in 21 cities (prefectures) and all counties; advances "5G + health" coordination; supports West China Hospital to build 5G smart medical public health collaboration platform; launches special actions to improve IPv6 end-to-end connectivity; and raises IPv6 traffic optimization capacity of data center, content delivery network, and cloud service platform. Sichuan continuously bolsters the development of information-based government services and e-government, and realizes the full coverage of WAN e-government of provincial departments. Sichuan sustainably executes "one-website handling", completes the basic connection between 40 provincial departmental business systems and provincial integrated government-service platforms, and consolidates the strength in cyberspace and the construction of "digital Sichuan".

Fujian takes efforts to implement digital economy innovation and development project, promotes digital industrialization and industrial digitalization, maintains fast growth in digital economy, and quickens the construction of national digital economy innovation and development pilot area. Meanwhile, Fujian deepens the construction of "digital Fujian", takes the lead in digital application, launches "Min Zheng Tong" government-service app, builds Diaoyu Dao Digital Museum of China, and firstly realizes the full-process application of administrative approval of e-certificates in China. In particular, Fujian constructs the first independent and open "city brain" in China, which provides more and better services for urban governance and public life.

Tianjin energetically promotes the construction of 5G and other new-type infrastructure, and basically achieves continuous external coverage in urban areas and key industrial application areas. Centering on smart city, intelligent manufacturing, and other key fields, Tianjin creates more use cases of 5G. Tianjin builds national high-level new-generation AI innovation and development pilot area; cloudifies industrial enterprises with digital core-technology ecology represented by Phytium, Kirin, and Sugon; and accelerates the construction of industrial Internet platforms. Tianjin vigorously supports the development of digital economy; strengthens the top-level design of digital economy; establishes and improves the policy system of industrial development, quickens industrialization and digitalization of data, introduces and cultivates industrial-platform enterprises, and fosters the integrated development of computing industry. Simultaneously, Tianjin continually reinforces the construction of digital government and smart city, and builds China-Singapore Tianjin Eco-City into a national benchmark smart city.

Hubei systematically deploys new-type infrastructure like 5G and industrial Internet, and builds modern industrial system dominated by advanced manufacturing industry and spurred by IT innovation. Hubei constantly promotes the development of "Chu Tian Yun" and basically forms provincial interconnected cloud service system, and realizes intensive sharing of government infrastructure resources such as computing, storage, disaster backup, service support, and security guarantee, which effectively shore up the construction of digital government in Hubei. Additionally, Hubei accelerates the construction of more than 40 provincial new-type infrastructure projects in industrial Internet top node, network trunk, big data center, 5G base station, industrial Internet platform, and digital industrial park; further advances the development of gigabit optical fiber, data center, IoT, and industrial Internet; and promotes the integrated development of digital infrastructure with intelligent manufacturing, smart health, smart transportation, smart logistics, smart city, and other fields.

The Rankings of the Second-Level Evaluation Indexes

In China Internet Development Index System in 2021, "Overview" Table 2 shows top ten provinces (autonomous regions and municipalities directly under the Central Government) in six aspects, i.e., information infrastructure construction, innovation capacity, digital economy development, Internet application, cybersecurity, and cyberspace governance.

1. The Ranking of Infrastructure Construction Index

Various regions attach great importance to network infrastructure construction; focus on optimizing and improving network quality; actively promote 5G commercial deployment and IPv6 scale deployment; and form broadband and mobile network infrastructure and service system that features high speed, urban-rural-area coverage, and convenient services. Among them, Guangdong, Beijing, Zhejiang, Shandong, Shanghai, Henan, Tianjin, Jiangsu, Liaoning, and Fujian rank top ten in China in infrastructure construction index. By April 2021, with 126,000 5G base stations and 69,000 NB-IOT base stations, Guangdong ranked the 1st in the number of national industrial Internet. In Beijing, the number of 5G user accounted for more than 21% of mobile phone user, and the number of 5G base station exceeded 42,000. Shanghai built more than 41,000 5G outdoor base stations and 66,000 indoor minor stations, and achieved 5G network coverage in 16 administrative areas, with the proportion of IPv6 active user reaching 59.9%.

2. The Ranking of Innovation Capacity Index

Various regions continuously invest capital and talent resources to support innovation and technological R & D, actively make breakthroughs in the innovation and R & D of new-generation information technology represented by AI and big data, and attain new progress in cutting-edge technologies. Among them, Guangdong, Beijing, Jiangsu, Zhejiang, Shanghai, Shandong, Hubei, Anhui, Tianjin, and Fujian rank top ten in China in innovation capacity index. Guangdong robustly supports technological R & D and innovation; totally invests 2.37 billion *yuan* to major special projects such as "virtual reality", "4K/8K UHD video", and "network collaborative manufacturing & industrial Internet"; attracts 4.63 billion *yuan* of social funds; develops 150 new products and 170 new technologies; formulates around 150 standards; and applies 2,000 items of invention and patent. This furnishes strong technological support for Guangdong to improve the development of Internet industry. In 2020, Beijing realized a business revenue of 1.77 trillion *yuan* in software and information service industry, with the scale of cloud computing, big data, AI, and other emerging fields all exceeding 100 billion *yuan*. Jiangsu introduces Yangtze River Delta Integrated Demonstration Zone International Innovation Center for Green Development to serve the integrated development strategy of the Yangtze River Delta; builds innovation cluster platform, demonstration platform, and international cooperation; and exchanges platform for green development; formulates various subsidy policies for talents and industrial parks; and creates a good environment for innovation.

3. The Ranking of Digital Economy Development Index

Various regions attach great importance to the development of digital economy, draw up master plans or action plans to promote the development of digital economy and digital industry, speed up digital industrialization and industrial digitalization, and achieve remarkable results in the informatization of manufacturing industry. Among them, Beijing, Zhejiang, Shanghai, Jiangsu, Guangdong, Shandong, Sichuan,

Henan, Fujian, and Chongqing rank top ten in China in digital economy development index. Beijing accelerates the construction of "Two Zones", i.e., National Service Industry Enlarging Opening-up Comprehensive Demonstration Zone and China (Beijing) Free Trade Pilot Area, makes plans for five tasks (i.e., the construction of international scientific and technological innovation center, the construction of "Two Zones", the construction of global digital economy benchmark city, the creation of new demands with supply-side structural reform and the relief of Beijing's functions not essential to its role as the capital), and builds a global digital economy benchmark city with spearheading standards. In 2020, Beijing realized the added value of digital economy of 1.44 trillion *yuan*, accounting for 40% of its GDP. Beijing attracts 1,500 AI corporations represented by Baidu and ByteDance, accounting for *c.* 28% of the total number of AI enterprise in China. Zhejiang encourages and guides enterprises to actively expand domestic market. In 2020, in above-designated-scale electronic information manufacturing industry, Zhejiang achieved an added value of 242.96 billion *yuan*, up by 16.8% year on year. Guangdong quickens the legislation in the field of digital economy, formulates *The Interim Measures for the Opening of Public Data in Guangdong Province*, and optimizes the construction of data-trading venues and supporting institutions, ranking the 1st in 4K user, industrial scale, the business income of above-designated-scale Internet enterprise, and the scale of digital economy in China.

4. The Ranking of Internet Application Index

Various regions highlight the development of Internet application. Internet-related industries maintain a trend of rapid growth. Internet deeply integrates with government service, education, and other public services and produces remarkable results. Among them, Beijing, Shanghai, Shandong, Guangdong, Sichuan, Chongqing, Fujian, Anhui, Zhejiang, and Tianjin rank top ten in China in Internet application index. Focusing on the vision of "getting information with one screen and managing the whole city with one website", Shanghai builds online and offline integrated service platform to handle 3,177 items of services, including administrative approval, public services, and other administrative power matters. Fujian deepens the construction of "digital Fujian", takes the lead in digital application, promotes the construction of digital culture and smart tourism, implements "Fujian Cultural Memory" Digital Culture Project, builds Diaoyu Dao Digital Museum of China, launches "Min Zheng Tong" government-service app, and firstly realizes the full-process application of administrative approval of e-certificates in China. In particular, Fujian constructs the first independent and open "city brain" in China, which provides more and better services for urban governance and public life.

5. The Ranking of Cybersecurity Index

Various regions actively optimize the construction of cybersecurity protection capacity, continuously improve cybersecurity guarantee capacity, rapidly promote the development of cybersecurity industry, organize diverse forms of cybersecurity publicity activities, and enhance public awareness of cybersecurity. Among them, Beijing, Guangdong, Shanghai, Shandong, Sichuan, Fujian, Jiangsu, Hubei, Chongqing, and Anhui rank top ten in China in cybersecurity index. Guangdong strengthens overall planning and coordination of cybersecurity work, continuously reinforces cybersecurity protection, consolidates the foundation of cybersecurity, hosts cybersecurity publicity weeks for consecutive years, and popularizes the laws and regulations like *The Cybersecurity Law of the People' Republic of China.* Noticeably, 230 million people take part in cybersecurity publicity week. On the basis of Shanghai Cybersecurity Trend Awareness Platform, Shanghai timely collects and analyzes cybersecurity information, and pools efforts in addressing cybersecurity issues.

6. The Ranking of Cyberspace Governance Index

Various regions lay great emphasis on cyberspace governance; actively introduce relevant governance regulations, industrial policies, and action plans; continually strengthen the top-level design of cyberspace governance; and improve the functions of cyberspace administration departments. Internet social organizations develop fast, which continue to promote the level of media convergence. Among them, Hebei, Shanghai, Beijing, Sichuan, Tianjin, Guangdong, Hubei, Shandong, Henan, and Shaanxi rank top ten in China in cyberspace governance index. Hebei further advances "Internet clean-up campaign in Hebei" Cyberspace Ecological Governance; investigates and tackles the problems of unhealthy cyberspace ecology, traffic on pornographic and vulgar information, celebrity gossip marketing, harmful social behaviors in cyberspace involving the minors and cyberspace violence according to law; and continuously purifies cyberspace. Besides, Hebei builds "Hebei Cyberspace Administration" omni-media matrix, gives play to the advantages and leading role of cyberspace administration accounts, and strengthens the production of Internet content and the dissemination of positive energy in a sustainable way. Focusing on the construction of "suggestions to leaders", "government-service microblogs", "government-service WeChat", and "client-end services", Sichuan basically establishes efficient and convenient Internet matrix service platform that connects three levels and different departments at the county, municipal, and provincial levels. Sichuan releases *Sichuan Internet Enterprises Integrity Declaration* to forge an honest-based cyberspace environment.

The Trends in China Internet Development

The "14th Five-Year Plan" signals the first-five-year period for China to embark on new journey of building a socialist modern country in all respects and realizing the second centenary goal after building a moderately prosperous society in all respects and achieving the first centenary goal. Specifically, in cyberspace development, relevant departments should build the synergy of various parties, strengthen the governance of Internet content, and accelerate the construction of comprehensive cyberspace governance system. They should quicken scientific and technological innovation, vigorously promote the breakthroughs in key core technologies in the information field, stress digital-technology-driven or guided development, advance digital industrialization and industrial digitalization, and inject new momentum to building new development pattern and achieving high-quality development. Additionally, they should prevent and address cybersecurity risks, reinforce cybersecurity guarantee system and capacity building in an all-round way, and cement new national cybersecurity mechanism. They should build a community with a shared future in cyberspace; enlarge international exchanges and cooperation in the field of cyberspace administration; promote the formulation of international rules on global cyberspace; and provide widespread Internet-public-opinion support, reliable cybersecurity guarantee, and strong information support for implementing the "14th Five-Year Plan" and building a socialist modern country in all respects.

Network Infrastructure Plays a Prominent Role and the Construction of New-Type Infrastructure Accelerates

Hammered by the COVID-19 pandemic, global economy faces the pressure of downturn, and infrastructure construction becomes an important driving force for economic development. It is an essential requirement to improve the construction of network infrastructure for tapping new driving forces and achieving high-quality economic development. Presently, the CPC Central Committee attaches great importance to the construction of network infrastructure, especially new-type infrastructure, and creates a healthy policy environment for strengthening the construction of new-type infrastructure, improving IT industrial infrastructure and industrial chain, and promoting the deep integration of new-generation information technology and manufacturing. Simultaneously, the huge demands that arise from the digital transformation of Chinese economy and society meant big space for the application of network infrastructure, and the construction of new-type infrastructure faces such challenges as the dependence on the import of core components and the shortage of talents.

In the future, China continues to boost the construction of "dual gigabit" network infrastructure, expand the coverage of gigabit optical-fiber network, accelerate the large-scale deployment of 5G Standalone (SA) network, strengthen the R & D of 6G standards and key technologies, and comprehensively promote the commercial deployment of IPv6. China will raise the level of new-type infrastructure construction in an all-round way, strengthen the planning and standard construction in the key fields of new-type infrastructure, optimize the layout of data center infrastructure construction, and expedite the construction of national integrated big data center collaboration and innovation system. Besides, China will consolidate the foundation of industrial Internet network system, facilitate industrial Internet network connectivity project, and foster the construction of "comprehensive + characteristic+ professional" industrial Internet platform system. China will deploy new-type network with global coverage like satellite communications network to continuously promote the industrialization and commercialization of BDS.

Digital Economy Maintains High-Quality Development and the Governance System Continually Improves

Under the strong leadership of the CPC Central Committee and the State Council, China's digital economy develops rapidly and makes remarkable achievements in enhancing economic resilience and boosting high-quality economic development, becoming a new driving force for the development of national economy. Meanwhile, in China, the development of digital economy is beset by weak basic technology and industrial system, behindhand innovation capacity and inadequate high-end supply, lagging behind European countries and the United States in the deep integration of digital technology with industry and agriculture. The homogeneity in the development path of regional digital economy intensifies the competition for resources and imbalanced regional development. The malpractices of digital platform like of "big data discriminatory pricing" and "one out of two" often occur, which pose new demands and challenges for digital economy governance.

In the future, China continues to increase investment to the R & D in the digital field; strengthen basic research; enhance the innovation in key core technologies; promote the deep integration of digital technologies with real economy; implement "cloud-based and digital-technology-empowered development"; expand emerging digital industries; accelerate the digital transformation of industry, agriculture, and service industries; and cultivate new momentum for high-quality development. China will uphold the coordinated advancement of local advantage and regional integration of digital economy, develop the industries in digital economy based on regional conditions, form digital industrial clusters with regional characteristics and international competitiveness, promote the experience and model of leading regions in digital economy, and enable backward regions to share digital dividend. China will give full play to the advantages of free trade zones (ports), build national digital service export bases, and establish internationally competitive digital trade and digital economy development demonstration zones. China will improve the system and standard construction in the digital filed, deepen the reform of "streamlining administration and delegate power, improving regulation and upgrading services" in digital economy, explore the regulatory system that agrees with the development trend of digital economy in China, and optimize the environment for digital economy development, in order to achieve the goal that the added value of core industries of digital economy accounts for 10% of GDP by 2025. China will optimize digital economy governance system; augment the research and governance of data ownership, anti-monopoly regulation, and digital tax collection; establish and improve the rules on data element market; enhance graded and classified protection of data; reinforce the regulation of Internet platform economy; and quicken the standardized, orderly, healthy, and sustainable development of digital economy and industry.

The Construction of Digital Society and Digital Government is Deeply Implemented and Information-Based Services for the Public Are Continuously Upgraded

The Outline of the 14th Five-Year Plan for National Economic and Social Development proposes to promote the construction of cyberpower; accelerate the construction of digital economy, digital society, and digital government; and drive the reform of production mode, lifestyle, and governance model with digital transformation. In recent years, China significantly improved the construction of digital society and digital government. Meanwhile, the imbalanced development in various regions became prominent, the supply of high-quality social resources remained insufficient, the services cannot meet the targeted needs of vulnerable groups like the elderly, and information islands existed, which affected the effectiveness of the construction of digital society and digital government.

In the future, China will carry out the planning and deployment of *The Outline of the 14th Five-Year Plan for National Economic and Social Development*; focus on major fields of education, medical care, elderly care, and employment; speed up the development of Internet hospital, smart library, and virtual museum; and expand the coverage of high-quality public service resources. China will optimize the functions of national integrated government-service platform, strengthen the opening and sharing of public data, realize the joint contribution and shared benefits of government information, raise the efficiency of digital government services, and build unified national opening public data platform, so that the masses can "handle affairs at the nearest service halls" and "handle affairs in a self-service way", market entities can experience the upgraded development environment with smart services, and information-based services for the public can complete "the last mile". China will accelerate the construction of smart city and digital village as well as smart community and digital family, promote the construction of smart community nationwide, narrow the digital divide between urban and rural areas, boost the construction of rural network infrastructure, improve the level of intelligent production and networked management in agriculture, and optimize rural digital governance system in a sustainable and progressive way. China will take into account the needs of digital vulnerable groups like the elderly, their concern of the design of relevant application products in particular. China will organize public welfare training for digital vulnerable groups with the help of communities, universities for the elderly and public welfare institutions, so that they can master basic Internet knowledge and enjoy the convenience of digital technology.

Innovation-Driven Development of Internet Content is Achieved and the Construction of Internet Civilization is Strengthened

Presently, information technologies like 5G, big data, cloud computing, IoT, blockchain, and AI lead and drive the development of media convergence. Internet communication channels are diversified, with a platform-based, intelligence-based and multi-function-based trend among the public in new media. Overall cyberspace governance produces remarkable results and cyberspace becomes increasingly clean. However, the spread of the false information on the pandemic, Information Cocoons, and Echo Chamber Effect intensify, with new problems and challenges emerging.

In the future, China will improve omni-media communication system; enhance the communication, influence, and credibility of mainstream media; spread positive energy on Internet; and propagate mainstream socialist values. China will use emerging technologies to develop interactive, service-oriented, and experiential information service forms, and optimize the professionalism, diversification, and intelligence of Internet content production and delivery. China will combine "Going Global" with "Introducing Overseas Creators", actively promote Internet cultural exchanges and mutual learning, strengthen the construction of international communication, accelerate the construction of Chinese discourse and narrative system, tell Chinese stories and spread Chinese voice. China will enhance the construction of Internet civilization, develop positive and healthy Internet culture, quicken the construction of comprehensive cyberspace governance system, improve technological capacity and system construction in cyberspace governance, change the way of regulation, and raise the effectiveness of cyberspace governance. China will strengthen minors' Internet protection, highlight the psychology of young Internet users, give full play to the role of various parties, promote the legislation on minors Internet protection, encourage and support the creation and dissemination of Internet content for the healthy growth of young people, and build a good environment for the growth of the minors. China will give full play to the power of Internet users, rely on Internet users closely, advance the construction of Internet illegal and harmful information reporting and supervision system, and use Internet joint rumor refutation platform to improve the quality of Internet users, guide Internet users to voluntarily restrict their Internet behaviors, and further purify cyberspace.

The Construction of Cybersecurity Guarantee System Is Accelerated and Data Security Protection Is Continuously Strengthened

Currently, emerging technologies like cloud computing, big data, IoT, mobile Internet, and AI develop fast, and new digital models and new business forms grow against the headwinds. The sharing and flow of data elements, the iterative development of emerging technologies, and the construction and application of new-type infrastructure aggravate cybersecurity risks and increase the difficulty of cybersecurity protection. However, the existing cybersecurity infrastructure remains relatively weak, core technologies rely on others, security protection system proves imperfect, and security testing standard and evaluation mechanism lags behind, with the shortage of cybersecurity personnel and the unsatisfactory construction of talent team. It is an urgent task to continually strengthen cybersecurity education and training and improve the awareness of cybersecurity among the public.

In the future, China will comprehensively strengthen the construction of cybersecurity guarantee system and capacity; prioritize safeguarding national data security; increase investment to critical information infrastructure and cybersecurity; solidify the construction of national cybersecurity attack and defense system; continuously improve the security defense capacity of critical infrastructure; integrate industry, university, and research forces in independent R & D innovation; collaborate with various parties in the R & D of basic technologies, joint address key technological problems; and enhance supply-chain security. China will strengthen the innovation and application of cutting-edge technologies in the field of cybersecurity, accelerate the iterative upgrading of cybersecurity products and services, and constantly meet diverse demands for security protection under new use cases. China will fortify the protection of personal information, implement the provisions of *The Personal Information Protection Law of the People's Republic of China*, and establish and improve the supporting systems and standards for the protection of personal information and data security. China will continue to improve the mechanism of cybersecurity talent training and the construction of cybersecurity discipline, and support universities and cybersecurity enterprises to jointly establish cybersecurity colleges, training bases, and cybersecurity laboratories. China will give play to the leading role of domestic and international skill competitions and cultivate high-level talents for cybersecurity.

International Exchanges and Cooperation in Cyberspace Continue to Deepen and Various Countries Play a Greater Role in Making International Rules on Cyberspace

Since the outbreak of the COVID-19 pandemic, telemedicine, online education, sharing platform, and office automation have been widely applied, with which data leakage and cyberattack occur. Intelligent manufacturing, energy-saving conservation and emission reduction, health, AI, and other emerging technologies are intertwined and integrated, posing new governance needs to international community. Simultaneously, China faces more serious and complex environment in the participation in international cyberspace governance. Several countries distort the achievements China made in digital economic development, constantly curb and contain the information technologies and products in China, agitate for the so-called "technology decoupling", and endanger the supply-chain security of China.

In the future, China will deepen international exchanges and cooperation in cyberspace; promote building a community with a shared future in cyberspace, advance the formulation of international rules in cyberspace with the United Nations as the main channel, establish multilateral, democratic and transparent global Internet governance system, actively participate in major events on international cyberspace governance platforms, and enhance communication and equal consultation on the basis of seeking common ground while shelving differences. China will host World Internet Conference in a productive way, build an international platform for the connectivity between China and the world as well as a Chinese platform for the joint contribution and shared benefits of international Internet. China will give full play to the role of governments, international organizations, Internet enterprises, social organizations, and think tanks; cling to the combination of high-level dialogues and non-governmental dialogues; and actively participate in the formulation of international rules and standards in the digital fields such as data security, digital currency, and digital tax. China will strengthen the exchanges and cooperation with the United States, Russia, the European Union, ASEAN, Africa, and other countries and regions; increase trust and clear doubts; deepen cyberspace exchanges and cooperation under the frameworks of Asia-Pacific Economic Cooperation, BRICS and Shanghai Cooperation Organization; and launch practical cooperation with BRI countries and regions.

The Ecology of Internet Public Welfare Takes Shape and the Regulation and Capacity Building Advance Side by Side

The vigorous development of Internet public welfare expands the boundaries of traditional public welfare industries, evokes the enthusiasm of the public to participate in public welfare, and gathers public welfare resources from various sectors. However, the existing environment and regulatory system of Internet-public-welfare system in China needs to be further constructed and improved, and the nature of personal help-seeking in Internet public welfare needs to be clarified. Illegal donation, fraud, and other acts that violate public welfare ethics occur from time to time, which undermine the credibility of Internet public welfare.

In the future, China will revise *The Charity Law of the People's Republic of China*, optimize supporting policies and measures, and build a favorable institutional environment for the development of Internet public welfare. China will clarify the qualification setting of public fundraising in Internet fundraising, the regulation on public-fundraising filing and the admission of public-fundraising platform, establish and improve administrative coordination mechanism and classification management system in charity, and formulate different regulatory policies for different types and scales of charitable organizations. China will strictly punish the acts of swindle, donation fraud, and misappropriation of charitable property, protect the legitimate rights and interests of donors, strengthen the whole-process all-round transparency of data and information of Internet public welfare, improve the public accessibility to data and information, and enhance the social credibility of Internet public welfare. Additionally, China will build Internet-public-welfare standard system; raise the ability of charitable organizations to operate charitable projects in a correct way; and promote their abilities in information collection, management, and disclosure.

Contents

Chapter 1
The Construction of Information Infrastructure

1.1 Outline

China energetically promotes the construction of information infrastructure and attains remarkable achievements in this regard. The construction of communication network infrastructure like 5G, optical-fiber networks, IPv6 and space-based networks is advancing steadily. China builds the world's largest 5G Standalone (SA) network and continuously improves technological R&D capacity and industrial level. The optical-fiber broadband network achieves full coverage in urban and rural areas and enters the Gigabit era for the first time. China makes significant progress in the large-scale deployment of IPv6 in basic network facilities, application-oriented infrastructure, terminal and industrial applications. The construction, demonstration application and industrialization of space-based network facilities continue to develop. The construction of computing infrastructure represented by data centers, cloud computing, edge computing and supercomputing accelerates. The market scale keeps expanding, and new models and applications continually emerge, which provide strong support for the digital transformation of the economy and society. The deployment of new technological infrastructure like AI and blockchain quickens, with industrial scale stably growing and the application field continuously enlarging. Application-oriented infrastructure like IoT, industrial Internet and IoV bolsters high-quality development in a faster and deeper way.

1.2 The Construction of Communication Network Infrastructure Continues to Accelerate

Spurred by the launch of a series of national policies and the wave of "new infrastructure", China continues to make breakthroughs in 5G-technology innovation, strengthen development momentum and stimulate industrial integration. China increases investment to the construction of optical-fiber networks and upgrades

Chinese Academy of Cyberspace Studies, *China Internet Development Report 2021*,
https://doi.org/10.1007/978-981-19-9311-4_1

network performance, computing capacity and platform construction. The popularization of IPv6 speeds up, and the number of active users grow quickly and sustainably. The space-based network system technology undergoes steady growth after starting from scratch.

1.2.1 China Takes the Lead in the World in 5G Commercial Development

1. The Largest 5G Network in the World Is Built

Presently, China has built the largest 5G Standalone (SA) network in the world. By June 2021, the number of 5G base stations totaled 961,000 in China, accounting for more than 70% of the world. China has achieved a full-coverage 5G network in cities at and above the prefectural level and has become a front-runner in global 5G development. The scale of terminal user continues to rise. By June 2021, the number of 5G terminal-connected users in China reached 365 million, with a net increase of 166 million over the end of 2020.[1] China's 5G network adopts the scale deployment of SA model. In November 2020, China Telecom firstly announced the scale commercial use of 5G SA network, planning to commercialize 5G SA network in more than 300 cities in China. In the same month, China Mobile also announced the scale commercial use of 5G SA network. Simultaneously, China Unicom accelerates the transition from 5G Non-SA network to 5G SA network. Predictably, by 2025, the investment in the construction of 5G network will amount to 1.2 trillion *yuan*, which drives the application-oriented investment to the upstream and downstream of industrial chains and various industries.[2]

2. Technological R&D and Industrial Strength Are Continually Enhanced

China maintains a global leading role in 5G technology. In the standard essential patent for 5G declared by European Telecommunications Standards Institute, the number of patents declared by Chinese enterprises accounts for 38%. Chinese enterprises continue to increase their investment in the R&D of 5G basic technologies and focus on the production & manufacturing of 5G terminals. In the first half of 2021, domestic shipments of 5G mobile phones reached 128 million, accounting for 73.4% of the world's. As estimated, in the second half of 2021, the shipment of 5G mobile phones would account for *c*. 80%. In the middle & high-end mobile phone market

[1] The State Council Information Office: "The Press Conference on the Development of Industry and Information Technology in the First Half of the Year Held by the State Council Information Office". July 16, 2021, http://www.scio.gov.cn/xwfbh/xwbfbh/wqfbh/44687/46299/index.htm.

[2] www.chinanews.com.cn: "China Academy of Information and Communications Technology (CAICT) Estimates That the Total Investment to 5G Network Will Reach 1.2 Trillion *Yuan* by 2025". March 4, 2020, https://www.chinanews.com/cj/2020/03-04/9113848.shtml.

(priced at 2,000 *yuan*), the sales of 5G mobile phone shares more than 90%.[3] With the extensive use of 5G network, 5G mobile terminals are increasingly accepted by consumers.

3. The Use Cases of Innovation and Integration Are Constantly Expanded

In the two-year commercial use of 5G, 5G integrates with various industries in a fast way, with use cases continually expanding. In the field of consumption, basic telecom corporations launched a series of new products, such as 5G video coloring ringback tone, ultra-HD video and cloud gaming. Technological integration of "5G + big data + AI + AR/VR" breeds new growth points in the niche consumer markets like game & entertainment and residential services and spurs the development of application-oriented industries. New business models and forms, e.g., "5G + automatic driving" and "5G + smart power grid", continue to emerge. Applications like 5G telemedicine and 5G smart medical care robot turn into reality. Typical forms of application, such as "5G + smart hospital", emergency rescue, public health and health care & elderly care, successively start. "5G + 4 K livestreaming", teleworking and teleteaching play an important role in the prevention & control of the COVID-19 pandemic and enterprise's resumption of work & production.

1.2.2 Optical-Fiber Broadband Enters the Gigabit Era

1. The Largest Optical-Fiber Network in the World Is Built

Presently, China has completed the turn from copper-cable-to-home to optical-fiber-to-home. With all cities at the prefectural level in China built into "optical network cities" (full coverage of optical-fiber network), the fixed broadband access rate in urban areas generally exceeds 100 MB/s, and gigabit broadband has become popular in several developed cities. Fixed broadband access rate in rural areas exceeds 12 MB/s, and more than 98% of administrative villages and poverty-stricken villages get access to optical-fiber broadband. The number of fixed-broadband-access users increases year by year. By May 2021, the number of fixed-broadband-access users of China Mobile, China Unicom and China Telecom totaled 505 million, with a net increase of 21.61 million over the end of 2020 (see Fig. 1.1). Specifically, the number of users with the fixed-broadband-access capacity of more than 100 MB/s exceeded 461 million, accounting for 91.3% of the total number, up by 1.4% over the end of 2020.

China attaches great importance to the construction of a gigabit optical-fiber network. In the *Report on the Work of the Government of 2021*, it is clearly stated

[3] www.cinic.org.cn: "The Shipment of 5G Mobile Phones in Chinese Market Exceeded 91 Million in the First Four Months". May 18, 2021, http://www.cinic.org.cn/xw/tjsj/1093686.html.

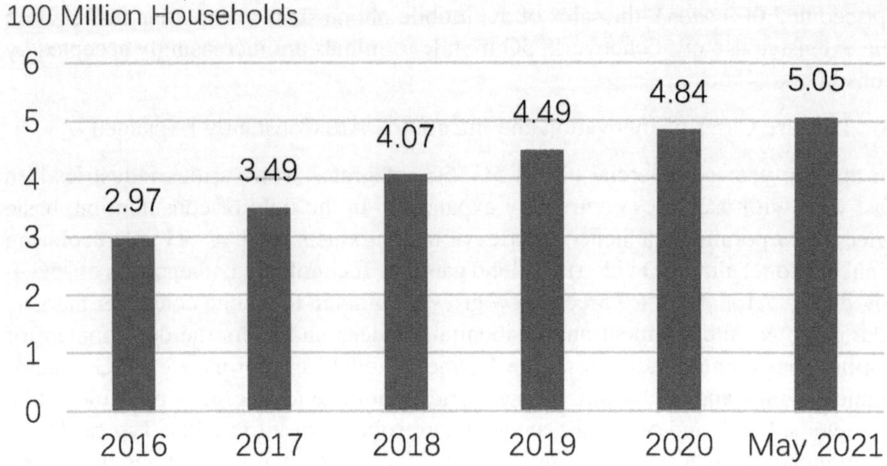

Fig. 1.1 The scale of fixed-broadband-access users in China from 2016 to May 2021 Data Source:Ministry of Industry and Information Technology

that China will strengthen the construction of 5G network and gigabit optical-fiber network and diversify their use cases. The "14th Five-Year Plan" also proposes to promote and upgrade gigabit optical-fiber network. In March 2021, the Ministry of Industry and Information Technology issued *The Action Plan for the Coordinated Development of "Double Gigabit" Networks 2021–2023*, which serves as the top-level design and guiding document for the development of China's broadband network in the new era. China vigorously boosts the development of gigabit optical-fiber networks, whose construction & scale remain on the top in the world. By May 2021, the number of users with gigabit-or-above access rate reached 12.19 million, with a net increase of 5.79 million over the end of 2020, accounting for 26.8% of a year-on-year net increase in fixed broadband access users. Currently, China has deployed gigabit broadband optical-fiber network in more than 300 cities, with the number of users of the gigabit optical-fiber network exceeding 80 million. Big cities like Beijing, Shenzhen, Shanghai, Hangzhou, Chengdu and Wuhan all set goals for the construction of gigabit broadband. In particular, China Mobile Beijing Branch is quickening the construction of 5G fixed network (F5G) and plans to build 1,000 double-gigabit high-quality communities in Beijing. China Telecom Guangzhou Branch and Huawei release the first *White Paper on FTTR* in the industrial circle, which aims to ensure the full-home coverage of gigabit broadband via home-all-fiber network solutions.

2. Gigabit Broadband Optical-Fiber Network Plays an Increasingly Prominent Role in Empowering Other Industries

The popularization of gigabit broadband optical-fiber networks improves the experience of broadband networks for the public as well as governmental and corporate customers. Besides, its technological advantages, such as ultra-high bandwidth, all-fiber interconnection and high stability, help to generate business innovation and incubate new use cases and models. At the home-level application, gigabit broadband optical-fiber network boosts the popularization and development of video & entertainment, online education, online healthcare, teleworking and smart home. At the business-level application, gigabit broadband optical-fiber network backs enterprise-targeted high-quality network, enterprise cloudification and all-fiber industrial park that develop fast, and facilitates the digital transformation of national pillar industries like transportation, electricity, oil & gas and finance. At the industrial-level application, industrial broadband optical-fiber network strengthens the capacity of industrial Internet, actualizes a stable, green and energy-saving network featuring high bandwidth and anti-electromagnetic interference, unclogs information channels in all links, and fosters the digitization of industrial production.

1.2.3 The Scale Deployment of IPv6 Advances Remarkably

1. IPv6-Based Upgrading and Transformation of Network Facilities Accelerate

IPv6-based upgrading and transformation of Long Term Evolution (LTE) network and fixed broadband network facility are completed in an all-round way. 5G construction shores up IPv6 simultaneously. Three major telecom operators, i.e., China Mobile, China Telecom and China Unicom, have allocated IPv6 addresses to 1.44 billion fixed and mobile terminals. IPv6 traffic in the LTE network accounts for 13.6%. In application-oriented infrastructure, super large, large, and medium & small data centers of three telecom operators as above-stated basically complete IPv6-based upgrading and transformation, and CDN (Content Delivery Network) nodes that support IPv6 exceed 92.67%. IPv6-based upgrading and transformation of cloud-service enterprises accelerate. 11 domestic cloud-service enterprises bolster IPv6-based upgrading and transformation of commonly-used cloud products, such as cloud host, container engine, load balancing, domain-name resolution and object storage. In website-related applications, top 100 commercial websites or applications in China all support IPv6 access. In the websites of governments, central State-owned corporations and news media, the support rate of IPv6 also considerably improves.

2. The Number of IPv6 Active User Increases Rapidly

In order to implement the tasks and requirements of *The Action Plan for Promoting the Scale Deployment of Internet Protocol Version 6 (IPv6)* released by the General Office of the CPC Central Committee and General Office of the State Council and the *Notice on Accelerating and Promoting the Scale Deployment and Application of Internet Protocol Version 6 (IPv6)* released by the Office of the Central

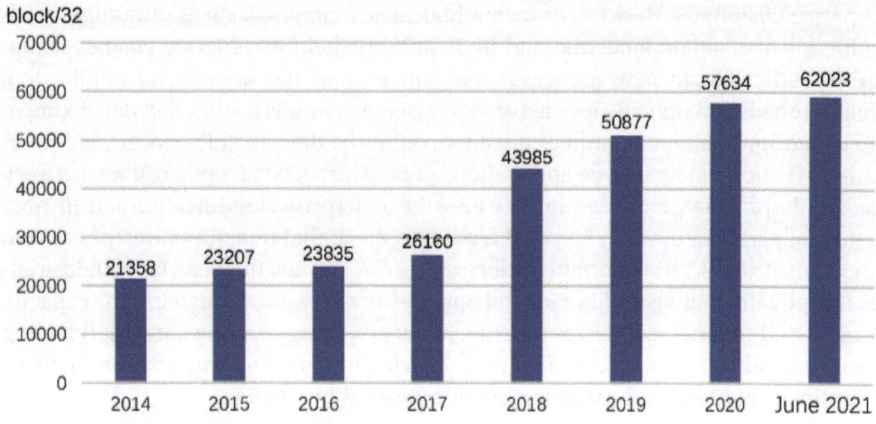

Fig. 1.2 The number of IPv6 address in China from 2014 to June 2021. Data Source: CNNIC

Cyberspace Affairs Commission, the National Development and Reform Commission and the Ministry of Industry and Information Technology, the Office of the Central Cyberspace Affairs Commission and the Ministry of Industry and Information Technology jointly issue the *Three-Year Special Action Plan for IPv6 Traffic Improvement 2021–2023* to promote the continuous improvement of IPv6 traffic scale and expedite the smooth evolution and upgrading of the Internet toward IPv6. By May 2021, the number of IPv6 active users in China had reached 528 million, the proportion of IPv6 Internet active users in China had realized 53.39%, the number of IPv6 terminal active connections in China had totaled 1.436 billion and achieved 69.03%. According to the data of China Internet Network Information Center (CNNIC), by July 2021, the number of IPv6 active users in China reached 535 million, and the number of IPv6 addresses realized 62,023/32 (see Fig. 1.2), with an increase of 7.6% over the end of 2020.

1.2.4 The Construction of Space-Based Network Facility Continues to Advance

1. BeiDou Navigation Satellite System Is Completed in an All-Round Way

Presently, China boasts the third country in the world that independently develops global satellite navigation system. In July 2020, BeiDou-3 Navigation Satellite System was officially launched, which signaled the successful fulfillment of the "Three-Step Development Strategy" for the construction of China's BDS as well as a new era for BDS global service. Now, BDS is comprehensively applied in transportation, disaster relief & mitigation, urban governance and other industries, integrated into national core infrastructure like electricity, finance and communications, and adopted in livelihood-related areas like automatic driving, e-commerce, mass consumption and sharing economy. According to the *2021 White Paper on the Development of China's Navigation Satellite and Location Service Industry* published by GNSS and LBS Association of China, in 2020, the total output value of China's navigation satellite and location service industry achieved 403.3 billion *yuan*, with a year-on-year increase of 16.9%. In 2020, domestic shipments of smart phones with satellite navigation & positioning functions reached 296 million. By 2035, China will build and improve a more ubiquitous, more integrated, more intelligent and safer BDS-based integrated positioning and navigation time-service system, further the development of satellite navigation and location service industry, and provide robust support for the digital and intelligent transformation of economy and society.

2. The Construction of Satellite Internet Infrastructure Is Staged

In April 2021, with the approval of the State Council, the China Satellite Network Group was established, which would mainly undertake the construction and development of China's satellite Internet. Its founding ushers in a new stage for the construction of China's satellite Internet infrastructure. Satellite Internet signifies an Internet broadband service model with communication satellite as the access means. Satellite Internet forms large-scale networking after having launched a number of satellites, via which it extends globally and builds a larger satellite system with real-time information processing capability. Satellite Internet means a new-type network that can provide communication service (e.g., broadband-Internet access) to ground, ocean and air terminals, featuring wide coverage, low delay and low cost. In April 2020, as information infrastructure, satellite Internet was included in key areas supported by the national "new-type infrastructure" policy for the first time. In July 2021, China successfully launched Tianlian-I-05 Satellite. After entering the orbit and completing the test, Satellite 05 will realize global networking & operation with Tianlian-I Satellites 01, 02, 03 and 04 and Tianlian-II-01 Satellite. This evidences that China has become the second country in the world with a relay satellite system with global-coverage capability. Since in orbit, Tianlian-I-05 Satellite has furnished space-based TT&C (telemetry, track and control) and data relay services for China's manned spaceflight, space station, low & medium-orbit spacecraft and launch vehicle. It serves as a vital hub for China's space information transmission and forges a solid foundation for the development of China's space-based TT&C system. On July 25, 2021, China succeeded in the first low-orbit broadband satellite & 5G private network integration test. The low-orbit broadband satellite was used to build the backbone network between 5G private networks in Beijing and Jinan, which replaced ground optical fiber to be laid in advance and realized the integration of the low-orbit satellite and 5G private network. In the future, the low-orbit broadband satellite can ensure wireless communications for remote areas and emergency communication cases. In August 2021, China's Long March 6 Launch Vehicle successfully launched two multimedia beta test A/B satellites named "Songjiang" and "G60" into the orbits, which propelled the development of the cause of China's satellite Internet.

3. The Application of Remote Sensing Satellite Speeds up

China's satellite-remote-sensing technology plays an indispensable role in major projects and tasks, such as land resources survey, Project of Natural Gas Transmission from West to East China, South-to-North Water Diversion, Three Gorges Hydroelectric Power Station, Three-River & Three-Lake Water Pollution Prevention, turning marginal farmland to forests, sand prevention & control, traffic planning & construction, coastal zone monitoring and island mapping. In March 2021, China's Gaofen-12-02 Satellite was successfully launched. Gaofen-12-02 Satellite is a microwave remote sensing satellite deployed by "CHEOS (China High-Resolution Earth Observation System) National Science and Technology Major Project". Its ground pixel resolution can reach the sub-meter level. In July 2021, China successfully launched ten-group remote sensing-30 satellites. In May 2021, HY-2-D Satellite was successfully launched at Jiuquan Satellite Launch Center. HY-2-D Satellite constitutes China's first marine dynamic environment monitoring network with HY-2-B Satellite and HY-2-C Satellite launched in 2018 and 2020 respectively. It significantly improves global efficiency and accuracy of marine monitoring (the coverage rate is up by more than 80%) and effectively serves China's marine disaster prevention & mitigation, marine transportation and marine-economy development. In July 2021, FY-3E, the first civilian dawn-dusk orbit meteorological satellite in the world, was successfully launched at Jiuquan Satellite Launch Center. FY-3E will be networked with in-orbit FY-3C and FY-3D, which vigorously builds China into the only country in the world with the observing capacity of three orbit-meteorological-satellite networks (i.e., morning, afternoon and dawn-dusk), and effectively improves the accuracy of global numerical weather prediction.

1.3 The Construction of Computing Infrastructure Becomes a Key Sector in New-Type Infrastructure Construction

In recent years, the construction of data centers in China enjoyed the biggest development opportunities. The computing industry develops rapidly, with an average annual growth rate of more than 30%. The scale of the cloud computing market is expanding fast, with a year-on-year increase of more than 33% over 2020. Edge computing has become a hot field in the Internet industry, and the demand for its application continues to rise. The construction of supercomputing centers enters the "fast lane", with more than ten supercomputing centers constructed or under construction.

1.3.1 The Layout of Data Centers Continues to Advance

1. China Formulates Policies to Strengthen the Construction and Layout of Data Centers

The "14th Five-Year Plan" clearly proposes to "accelerate the construction of national integrated big data center system, enhance overall computing-power-based planning and intelligent coordination, build many national hub nodes and big data center clusters, and form E-level and 10E-level supercomputing centers". In December 2020, the National Development and Reform Commission, the Office of the Central Cyberspace Affairs Commission, the Ministry of Industry and Information Technology and the National Energy Administration jointly issued the *Guiding Opinions on Accelerating the Construction of National Integrated Big Data Center Collaborative Innovation System*, which proposed to "strengthen the top-level design of national integrated big data centers, deepen the collaborative innovation of big data in social governance and public service, finance, energy, transportation, commerce, industrial manufacturing and other fields, and develop regional data center clusters", to "forge a well-organized, green and intensive integrated infrastructure landscape of data center in China by 2025" and to "achieve a structural balance between data centers in Eastern China and Western China and reduce the electricity utilization rate of large and super-large data centers to 1.3 or below". In May 2021, 14 departments, including the National Development and Reform Commission, the Office of the Central Cyberspace Affairs Commission, Ministry of Industry and Information Technology and National Energy Administration, jointly released the *National Integrated Big Data Center Collaborative Innovation System Computing Hub Implementation Plan*, which clearly proposed to deploy and construct hub nodes of national computing network, start "Channeling Computing Resources from the East to the West" Project, and construct national computing network system. In July 2021, the Ministry of Industry and Information Technology issued *The Three-Year Action Plan for the Development of New-Type Data Centers 2021–2023*. The *Action Plan* clarifies that China will basically establish a well-organized development landscape of a new-type data center that features advanced technology and green and low-carbon development, with computing scale in alignment with the growth of the digital economy in three years.

2. The Scale of the Data Center Market Continues to Enlarge

By the end of 2020, in China, the scale of the data center market reached 195.8 billion *yuan*, with an increase of 39.5 billion *yuan* over 2019, up by 25.27% year on year. Predictably, the scale will exceed 590 billion *yuan* by 2025. Data centers are mainly distributed in economically developed and densely-populated first-tier cities like Beijing and Shanghai, with a trend of shifting to central and western regions. By the end of 2020, in the number of the framework of data centers, Beijing (and its surrounding areas), Shanghai (and its surrounding areas) and the western region rank top three in China, reaching 538,000, 520,000 and 369,000 respectively. Telecom operators have become the mainstay in the construction of data centers. Three major telecom operators run more than 60% of data centers. In particular, China Telecom accounts for the highest proportion, followed by China Unicom and China Mobile.

1.3.2 The Application Scope of Cloud Computing Continues to Expand

In 2020, in China, the scale of the cloud computing market achieved 177.64 billion *yuan*, with an increase of 33.41% over 2019. As estimated, the market scale would maintain the same growth rate in 2021, reaching 233.06 billion *yuan*. In the future, the cloud computing market in China will continue to develop at high speed. Now, the public cloud market outstrips the private cloud market, and becomes the first major market for cloud computing in China, whose market scale will exceed 230 billion *yuan* by 2023, as predicted. The scale of private cloud expands steadily, with a growth rate of *c.* 22%. By 2023, the scale of the private cloud market will approximate to 150 billion *yuan* (see Fig. 1.3). Besides, IaaS, PaaS and SaaS maintain rapid growth. In the "14th Five-Year Plan", the scale of IaaS market keeps rising, and the corporate demand for digital transformation drives the sustainable growth of market demand for PaaS like databases, middleware and microservices. As predicted, the scale of the PaaS market will maintain a high growth rate in the next decade. The SaaS of the public cloud turns to reality in a fast way and embodies great development potential.

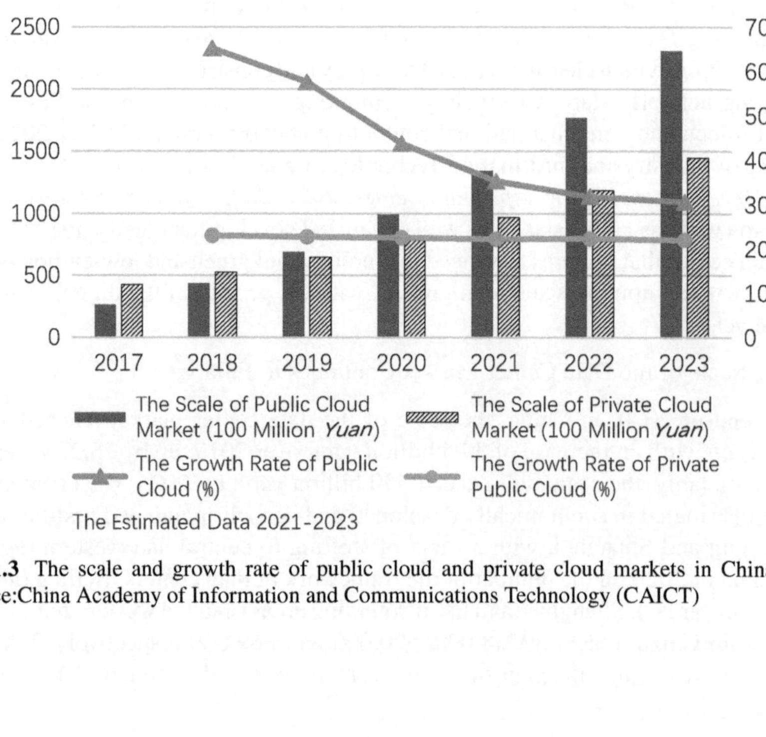

Fig. 1.3 The scale and growth rate of public cloud and private cloud markets in China. Data Source:China Academy of Information and Communications Technology (CAICT)

1.3.3 Edge Computing Possesses Tremendous Development Potential

1. Relevant Policies Boost the Development of Edge Computing

The Chinese government attaches great importance to edge computing and lays down many relevant policies to vigorously promote the application of edge computing to industrial Internet, intelligent manufacturing and other fields. In January 2021, the Ministry of Industry and Information Technology released the *Industrial Internet Innovation and Development Action Plan (2021–2023)*, which aimed to support industrial enterprises to comprehensively use technologies such as 5G, time-sensitive networking (TSN) and edge computing and improve the networking level of all production links. In May 2021, National Development and Reform Commission and three other departments jointly published the *National Integrated Big Data Center Collaborative Innovation System Computing Hub Implementation Plan.* As clearly stated, China will support the development of edge data centers, reserve development space for edge data centers with moderate scale and extremely-low-delay requirements in urban areas, encourage to take data centers in urban areas as the "edge" end of computing power, and give priority to meeting the needs of business with a high real-time requirement, such as high-frequency transactions in the financial market, virtual reality & augmented reality(VR/AR), UHD video, IoV, networked UAV, smart power, smart factory and smart security protection.

2. The Commercial Use of Edge Computing Emerges Rapidly

With the large-scale commercial use of 5G and the acceleration of new-type infrastructure, edge computing progresses from theoretical research to commercial use, and becomes inelastic demand in the digital transformation and development of vertical industries. Presently, 3.37% of Chinese enterprises in China have applied edge computing, and 44.23% of them plan to use edge computing. In the second half of 2020, the overall scale of China's edge computing server market achieved 1.542 billion U.S. dollars, which realized 2.655 billion U. S. dollars in the year, with a year-on-year increase of 16.3%. Specifically, the scale of the telecom network and industrial edge application market reached 274 million U.S. dollars, with a year-on-year increase of 129.6%. And general-purpose servers and edge customized servers accounted for 87.1% and 12.9%, respectively.[4] In the future, the edge-cloud collaboration presents the main development direction of edge computing capacity building.

[4] Data Source: *PRC Edge Computing Server Tracker in the Second Half of 2020* released by International Data Corporation (IDC) in April 2021.

1.3.4 The Construction of Supercomputing Center Enters the Fast Lane

By the end of 2020, more than ten supercomputing centers had been built or were under construction in China. In September 2020, Chengdu Supercomputing Center was completed and put into operation, whose application field covered aerospace, electronic information, biomedicine, equipment manufacturing, advanced material and energy & chemical engineering. In November 2020, Zhengzhou Supercomputing Center passed the acceptance. It provides strong computing support for the socio-economic development of Henan Province and its surrounding areas, and improves the level and capacity of national major information infrastructure in the central region. In December 2020, the construction project of the National Supercomputing Kunshan Center successfully passed acceptance. It focuses on applied computing research & service in frontier science such as AI, biomedicine, physical & chemical material and atmospheric & marine environment.

1.4 New Technological Infrastructure Creates New Engine for Socio-Economic Development

The integration and innovation of digital technologies like AI, 5G, cloud computing and big data (or understood as new-type infrastructure that empowers the development of various industries) provide underlying support for the development of a digital economy and industrial digital transformation. Blockchain technology and industry develop dramatically, whose relevant applications expand to multiple fields such as digital finance, IoT, intelligent manufacturing, supply-chain management and digital-asset transaction. Currently, quantum-communication-related technologies are making significant progress, which may realize large-scale application in many industries.

1.4.1 AI Plays a Key Role in Supporting Intelligent Development

1. The Scale of the AI Industry Continuously Expands

As AI technology penetrates into various vertical fields in a faster way, more industries launch intelligent upgrading. As estimated, by the end of 2021, the scale of the AI industry in China would reach 203.5 billion *yuan*, with a growth rate of *c.* 30%. Spurred by AI, the scale of integrated-solution-service for the industrial-application market will realize 1,280 billion *yuan*, and the scale of the intelligent governance market will reach 133.8 billion *yuan*, with an annual growth rate of *c*. 20%.[5] By the end of 2020, the number of Chinese AI enterprises totaled 6,425, with a year-on-year increase of 25.37%, accounting for 24.66% of the world, and ranking the 2nd in the world (only second to the United States, 38.3%).[6] In the layout of AI enterprises, more than 80% are distributed in three regions, i.e., Beijing-Tianjin-Hebei, Jiangsu-Zhejiang-Shanghai and Guangdong-Hong Kong-Macao Greater Bay Area. Noticeably, Beijing, Shanghai and Shenzhen own multiple advantages such as abundant talent resources, tremendous scientific & technological strength, rich innovation on use case and distinctive industrial cluster effect. In the three regions, the number of AI enterprise exceeds 1,000, which serve as an important engine for the development of the AI industry in China.

2. The Use Case of AI Continually Diversifies

Presently, China takes the lead in AI applications in the world. AI application not only plays an important part in new economy, new business model and new lifestyle but also makes a great contribution to promoting the transformation and upgrading of traditional industries. In 2020, AI technology played a prominent role in the prevention & control of the COVID-19 pandemic. It broke the bottleneck in staff flow, logistics, information flow and capital flow in various industries, prevented or addressed major public security risks, and facilitated the resumption of work & production of manufacturing enterprises. In the medical field, by the end of November 2020, ten AI products of nine enterprises in China obtained Medical Device Trading Permit Type III, which made a great breakthrough. In April 2021, the 4th Digital China Summit · Social Governance in the Context of Artificial Intelligence & Social Experiment Forum was hosted by the Cyberspace Administration of China and co-staged by Zhejiang Lab and The Institute for AI International Governance of Tsinghua University (I-AIIG). Under the theme of "AI Governance: AI Empowers Social Governance", the Forum gropes for the laws and experiences concerning how AI Empowers social governance, and probes into legal, ethical and social problems that come with the phased development of AI. In the "14th Five-Year Plan", under the guidance of national industrial policy, the AI industry comes to the stage of fast development. Competitive industries like intelligent chips, intelligent drones, intelligent connected vehicles and intelligent robots further develop, and AI industrial ecology matures. Deep use cases of AI in key fields, such as health care, finance, supply chain, manufacturing, home furnishing and transportation, continue to emerge.

---

[5] www.askci.com: "The Status Quo and Prospect of AI Industry Market in 2021". December 2020, https://www.askci.com/news/chanye/20201221/0842491317385.shtml.

[6] Data Source: *2021 White Paper on Artificial Intelligence Development* published by Shenzhen Artificial Intelligence Industrial Association.

1.4.2 The Deployment and Application of Blockchain Accelerate

1. National and Local Support Continuously Increases

In April 2020, National Development and Reform Commission clarified the scope of new-type infrastructure for the first time and took blockchain as an integral part of the new-technology infrastructure, which injected new vitality into the deployment and construction of blockchain-technology infrastructure and underlying platform in China. The *Outline of the 14th Five-Year Plan for National Economic and Social Development and the Long-Range Objectives Through the Year 2035* regards blockchain as one of emerging digital industries and raises the requirements like "focusing on alliance chain to develop blockchain-service platforms and application programs in the fields of Fintech, supply chain finance and government services". In June 2021, the Ministry of Industry and Information Technology and the Office of the Central Cyberspace Affairs Commission jointly released the *Guiding Opinions on Accelerating the Application of Blockchain Technology and Blockchain Industrial Development*. It proposes that by 2025, the blockchain industry in China will take an initial shape and reach the world's advanced level in comprehensive strength. Local governments take the initiative in introducing supporting policies to expedite the deployment of the blockchain industry. By December 2020, local governments of various provinces (autonomous regions and municipalities directly under the Central Government) in China had promulgated 463 supporting policies to sustain the development of the blockchain industry.[7] In the deployment of the blockchain industry, local governments forge their unique features. For example, Hunan Province highlights "industrial Internet + blockchain". Hainan Province explores the application of blockchain in tourism consumption. Jiangsu Province deploys bill blockchain platform in banking, insurance, leasing and other industries. Simultaneously, local governments stress the development and technological innovation of blockchain enterprises via blockchain industrial parks, actively incubate or introduce blockchain enterprises, and build a distinctive path for blockchain development.

2. The Development of Blockchain Infrastructure Quickens

As an infrastructure that provides blockchain-technology supporting service, the blockchain public service platform aims to help users create, manage and maintain blockchain networks and applications. By embedding blockchain-technology architecture into cloud computing platform and utilizing the advantage of cloud service infrastructure, the blockchain public service platform lowers development and use costs and features fast deployment, quick access and high security and reliability. Presently, blockchain public service platforms in China are mainly built by industrial leading enterprises or industrial organizations, typified by Huawei Cloud BCS Platform, Tencent Blockchain TBaaS Platform, Ali Cloud BaaS Platform and

[7] Data Source: *White Paper on China's Blockchain Industrial Development 2020–2021* published by China Center for Information Industry Development (CCID).

Baidu Blockchain Open BaaS Platform. With diverse use cases, the alliance chain displays a trend of fast development and plays a leading role in the development of the blockchain industry in China. New technologies like inter-blockchain, sidechain, multichain, shard technology, consensus algorithm and privacy protection continue to emerge.

3. The Application Fields and Use Cases Continually Expand

The use case of blockchain keeps expanding, from single digital currency to the financial industry, copyright protection, traceability & anti-counterfeiting, energy industry, sharing economy and IoT, which strengthens socio-economic development to some degree. By November 2020, blockchain-application projects in the financial field numbered the 1st in domestic filed blockchain-information services, accounting for 36%.[8] In 2020, blockchain technology was quickly applied and many innovative use cases were created, owing to the demands of the prevention & control of the COVID-19 pandemic, the promotion of enterprise's resumption of work and production, and the guarantee of market supply. In governmental affairs and services for people's livelihood, blockchain technology-based digital-identity contracts and data-storage services effectively ensure the security and authorized use of "identity health code" and personnel data. In the financial field, the State Information Center, China Mobile and China UnionPay co-organized the Blockchain-Based Service Network (BSN), the first national alliance chain application. In April 2020, BSN was officially launched for commercial operation, with the attempt to establish a trusted, controllable and scalable alliance chain for industrial, corporate and governmental applications.

1.4.3 Quantum Communication Makes New Progress

In recent years, major economies in the world attached great attention to the research and application of quantum technology represented by quantum computing, quantum communication and quantum measurement, equated quantum technology with national strategy and increased investment in quantum technology, in a bid to forge dominant position in international competition on quantum technology. China starts later in the research of quantum technology. Yet, by dint of strong support of policies and great efforts of boffins, China successfully achieves breakthroughs in the field of quantum communication, especially in the construction of metropolitan quantum communication networks, quantum commercial trunk lines and quantum satellites. Presently, China has completed Quantum Communication "Beijing-Shanghai Trunk Line", launched the "Micius" Quantum Communication Satellite, and realized the first intercontinental quantum communication in the world.

[8] Data Source: *Blockchain White Paper 2020* published by China Academy of Information and Communications Technology (CAICT).

1.5 Application-Oriented Infrastructure Props up High-Quality Development in a Faster Way

As a key infrastructure that supports the digital economy, IoT is applied in all aspects of socio-economic life, such as industry, agriculture, environment, transportation, logistics, security protection, home furnishing, health care, education, finance and tourism. To vigorously develop IoT possesses great significance for promoting the development of the national economy and social informatization and improving the level of social management and public service. As the product of deep integration of new-generation information technology and manufacturing industry, industrial Internet builds new-type production, manufacturing and service system with all factors, all industrial chains, all value chains and all-round connections via the all-side interconnection of people, machines and things. In a way, the industrial Internet serves as the key comprehensive information infrastructure for the realization of digital and intelligent transformation of the manufacturing industry. Catalyzed by national policies as well as the strategic deployment of new-type infrastructure, China's IoV leads the list in the world in technological innovation, application demonstration and industrial-ecology construction, with industrial scale constantly expanding.

1.5.1 IoT Industrial Ecology Improves Remarkably

1. IoT Maintains Rapid Development and Industrial IoT Embodies Great Potential

By 2020, the scale of the IoT industry in China exceeded 1.7 trillion *yuan*, ranking the top in the number of IoT connections in the world. According to the data released by the Ministry of Industry and Information Technology, the scale of China's cellular-IoT users continues to expand. By the end of May 2021, three telecom operators, i.e., China Mobile, China Telecom and China Unicom, had 1.258 billion users on cellular IoT terminals, with a net increase of 122 million over the end of 2020 (see Fig. 1.4). Specifically, in the terminals applied to intelligent manufacturing, smart transportation and smart public service, the users accounted for 17.3, 17.6 and 22.1% respectively. The number of smart-public-service terminal users increased by 21% year on year, with the most prominent growth.

2. Independent-Technological-Standard-System-Supported Industrial Ecology Basically Takes Shape

China initially established a relatively complete industrial chain covering software, hardware, network operation and IoT service. In the spatial distribution of the IoT industrial chain, industrial alliances and R&D centers are established in eight major cities, i.e., Beijing, Shanghai, Wuxi, Hangzhou, Guangzhou, Shenzhen, Wuhan and Chongqing. This forms four IoT industrial cluster development zones in the Yangtze River Delta, the Pearl River Delta, Circum-Bohai Sea Economic Zone and central &

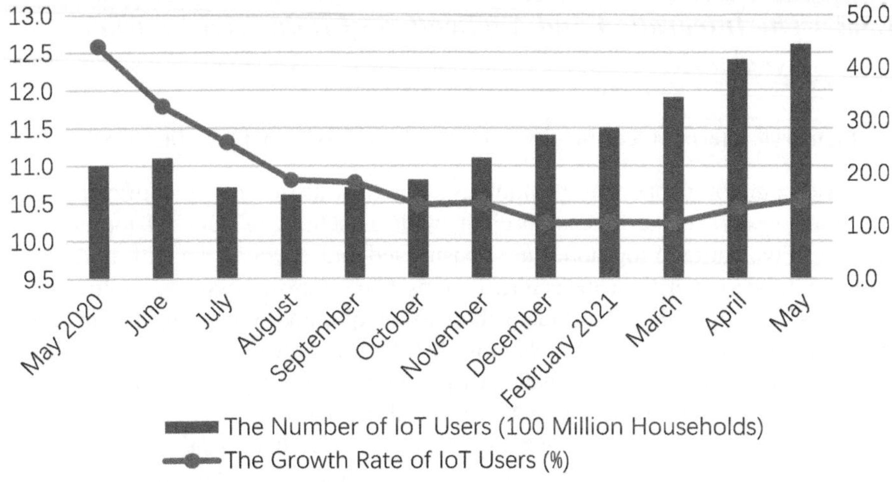

Fig. 1.4 The change in the number of IoT terminal users in China from May 2020 to May 2021

western regions, which extend to the surrounding areas. By the end of 2020, China Telecom had built the first NB-IoT commercial network in the world with continuous coverage, the largest scale, the widest coverage and the best frequency band, and deployed more than 410,000 NB-IoT base stations. Currently, China Telecom has more than 100 million 5G NB-IoT users and has become the first operator in the world with more than 100 million 5G NB IOT users,[9] ranking the 1st in the world in the number of NB-IoT connections. China Mobile has built more than 400,000 NB-IoT base stations and completed the whole-network upgrading based on 3GPP R14 standard, with the number of users in medium & low-speed IoT totaling 64.68 million.[10] The IoT of China Unicom has the largest single-connection management platform in the world, with 228 million IoT connections.[11]

[9] *People's Post and Telecommunications News*: "China Telecom Becomes the World's Largest NB-IoT Operator". May 17, 2021, http://www.cnii.com.cn/rmydb/202105/t20210517_278374.html.

[10] www.cnii.com.cn: "'Internet of Everything for an Intelligent Future: China Mobile's First IoT Medium & Low Speed Business Promotion Conference in 2021' Was Successfully Held in Guangdong". May 28, 2021, http://www.cnii.com.cn/tx/202105/t20210528_281959.html.

[11] www.xinhuanet.com: "'5G IoT for a Smart Era': China Unicom IoT Appears in 2020 Beijing Communications Exhibition". October 15, 2020, http://www.xinhuanet.com/info/2020-10/15/c_139442612.htm.

1.5.2 The Integration and Application of Industrial Internet Accelerate

1. Industrial Internet Identification and Resolution System Takes Initial Shape

As basic common supporting technology of industrial Internet, identification and resolution system make a breakthrough, with a toehold in national top-node as the core. Five national top nodes are constructed and operated steadily in eastern, southern, western, northern and central regions. China has successively deployed five top nodes in Beijing, Shanghai, Guangzhou, Chongqing and Wuhan, and two disaster-recovery nodes in Guiyang and Nanjing. By the first half of 2021, 156 secondary nodes had been launched, which were distributed in 25 provinces and cities. Covering more than 30 major categories of the national economy and providing access to around 20,000 enterprises, the number of secondary-node identification had exceeded 26 billion.[12] Simultaneously, taking distributed identification and blockchain as basic technologies, the construction of a decentralized new-type identification system quickens. Industrial Internet identification is applied in equipment manufacturing, automobile manufacturing, shipbuilding, construction machinery, food processing and other industries in a deeper way, with typical application models like product traceability, supply-chain management and lifecycle management. Besides, industrial Internet identification successfully connects terminals like IoT-device payment terminals and identification reading & writing device-terminal. The carrier and application of active identification are firstly deployed in the fields of instrumentation, industrial control and IoV in a large scale. This continuously improves the service capacity of public applications, reduces the enterprise's costs and enhances the enterprise's quality and efficiency to some degree.

2. The Scale of the Industrial Internet Market Quickly Enlarges

In 2020, the overall economic scale of the industrial Internet industry reached around 3.1 trillion *yuan*, with a year-on-year growth of *c.* 47.9%. Driven by the development of new-type infrastructure and the support of digital-economy policy, in the "14th Five-Year Plan", China's industrial Internet continues to maintain high-speed growth. By 2025, the overall scale of China's industrial Internet economy will achieve around 9.42 trillion *yuan*. To be specific, the added value of core industry of industrial Internet will reach 652 billion *yuan*, with a year-on-year increase of

[12] Lu Chuncong: "Industrial Internet Identification Registration Exceeds 26 Billion". https://baijiahao.baidu.com/s?id=1705227105578656077&wfr=spider&for=pc.

21.6%. From 2017 to 2020, the added-value scale of the core industry of industrial Internet rose by more than 93%, with a CAGR of 24.6%. By 2025, the added-value scale of the core industry of industrial Internet will amount to 1.96 trillion *yuan*.[13]

3. The Integration & Application Level of Industrial Internet Continuously Rises

With the large-scale commercialization and deployment of 5G network, the integration and innovation of "5G + industrial Internet" become more active, and the industrial ecosystem takes shape. With regard to four new models (i.e., intelligent production, networked collaboration, service-oriented expansion and large-scale customization), the number of enterprises that implement them accounts for 33.2, 26.8, 14.4 and 8.7% respectively.[14] Now, the use case of "5G + industrial Internet" mainly covers collaborative design, simulation verification, auxiliary assembly, precise control, flexible processing, nondestructive testing, remote maintenance and unmanned inspection. As the integration and application of "5G + industrial Internet" continues to deepen, industrial Internet gradually extends from external production links like patrol inspection and monitoring to internal or core production links like production control and quality inspection. Presently, China has formed "5G + industrial Internet" industrial clusters in the Yangtze River Delta, Guangdong-Hong Kong-Macao Greater Bay Area, Beijing-Tianjin-Hebei, traditional industrial bases in northeastern China and Chengdu-Chongqing Economic Circle. Meanwhile, the integration and application of "5G + industrial Internet" release more potential, and the integration and development of leading enterprises and small & medium-sized enterprises continuously deepen, which tremendously promote the cooperation between the upstream & downstream industrial chains.

1.5.3 IoV Market Scale Expands Fast

1. The Construction of the IoV Pilot Area Continues to Accelerate

Since 2020, the construction of the national IoV pilot area has been accelerating. In October 2020, the Ministry of Industry and Information Technology officially approved and supported Hunan Province (Changsha) in building the national IoV pilot area, which became the third MIIT-approved national IoV pilot area after Jiangsu Province (Wuxi, approved in May 2019) and Tianjin (Xiqing, approved in December 2019). In January 2021, the Ministry of Industry and Information Technology officially replied to Chongqing municipal government, supporting Chongqing (Liangjiang New Area) to build the national IoV pilot area. In September 2020, Beijing Economic-Technological Development Area (Beijing E-Town) announced to build the first Internet-connected cloud-controlled high-level automatic driving

[13] Data Source: *Industrial Internet Industrial Economic Development Report 2020* published by China Academy of Information and Communications Technology (CAICT).

[14] Data Source: *China Industrial Internet Development Effectiveness Evaluation Report* published by China Academy of Information and Communications Technology (CAICT).

demonstration zone in the world by the end of 2020, so as to support the large-scale operation of L4 and above high-level automatic driving vehicles. As planned, Beijing E-Town will complete the construction of five systems in 2022, i.e., "smart road, intelligent car, real-time cloud, reliable network and accurate map", which connect key links of the technology and management in relation to Internet-connected cloud-controlled automatic driving, form a city-level engineering test platform, and eventually realize high-level use cases such as unmanned expressway logistics, L4-level automatic driving taxi, intelligent Internet-connected bus and independent valet parking.

2. The Integration and Application of IoV Continue to Deepen and IoV Industrial Ecology Takes Shape

The strong support of national policies creates favorable opportunities for the development of the IoV industry and attracts voluntary participation of all parties in the industrial chain. ICT enterprises, application developers, service providers and content providers in the downstream industrial chain, vehicle manufacturers and terminal equipment manufacturers in the upstream industrial chain, and network operators that provide network connections for the entire industry, all make great efforts to develop the IoV industry. Presently, traditional Chinese automobile manufacturers, e.g., FAW, Chang'an Automobile, SAIC Motor, BAIC Motor, GEELY and Great Wall Motor, have carried out the R&D and testing of L1-L3 automatic driving. Baidu and FAW Hong Qi jointly develop Robo taxi, a Hong Qi EV automatic-driving taxi fleet, which launches a trial operation in Changsha. Mobile phone manufacturers like Huawei, Xiaomi, OPPO and vivo make a foray into automatic driving in a diverse way, by providing OEMs with smart car solutions, incremental parts and open-source software. In the application and promotion of IoV, two fields of logistics transportation and distribution service, develop fastest. Low-speed unmanned vehicles in limited areas are expected to be firstly commercialized.

Chapter 2
The Development of Information Technology

2.1 Outline

New-round scientific & technological revolution and industrial transformation progress vigorously. In particular, IT integration and innovation breed a variety of new models and new business forms, which reconstruct the global scientific & technological innovation landscape and reshape the global economic pattern. China adheres to high-quality development driven by scientific & technological innovation, continues to strengthen the innovation capability of new-generation information technology, and speeds up the construction of new development patterns in which domestic and foreign markets boost each other, with the domestic market as the mainstay.

Over the past year, China accumulated resources in original innovation on basic technologies like high-performance computing and software. With the cultivation of Open Source Ecology, innovation subject grows quickly. With multifaceted achievements made in cutting-edge fields like AI, cloud computing and industrial Internet, technological innovation and practical application form a positive cycle. In the field of new technologies and applications, cross-filed crossover technologies continuously integrate and develop. Applications like automatic driving and vaccine and drug R & D turn into reality and meet the urgent needs of society.

2.2 Basic Technologies

In 2020, high-performance computing developed rapidly, and E-class supercomputing and characteristic application advanced steadily. In software technology, some OS products came into service, and industrial software became one of major strategic goals. Open Source Ecology took shape, and open source products diversified, with breakthroughs made in key projects.

Chinese Academy of Cyberspace Studies, *China Internet Development Report 2021*,
https://doi.org/10.1007/978-981-19-9311-4_2

2.2.1 *With Breakthroughs Made in the Segmented Field of High-Performance Computing, Characteristic Applications Arouse Wide Attention*

High-performance computing plays a fundamental role in national scientific research and constitutes a source of scientific & technological innovation. Over the past year, overall performance and global ranking of supercomputing in China maintained stable levels, with significant progress in the research of graph processing, characteristic applications and other segmented fields. Relevant characteristic applications gained a foothold internationally.

1. Overall Ranking of Supercomputing Remains Stable

In June 2021, *The 57th Top 500 Supercomputer List* was released. Japanese supercomputer "Fugaku" ranked the 1st in a row, and American supercomputers "Summit" and "Sierra" ranked 2nd and 3rd respectively. Chinese supercomputers "Sunway TaihuLight" and "Tianhe-2A" ranked 4th and 7th respectively.

2. Segmented Fields Achieve More Breakthroughs

In June 2021, *Graph 500*, a list that measured the computing performance of super nomograph, was published. "Tianhe" Supercomputer, developed by the National University of Defense Technology and deployed at the National SuperComputer Center in Tianjin, ranked the 1st in *SSSP Graph 500* and *BIG Data Green Graph 500*, which demonstrated China's remarkable progress in AI and big data.

3. Characteristic Applications Emerge

In November 2020, the Institute of Applied Physics and Computational Mathematics (IAPCM), in collaboration with UC Berkeley of the United States, realized the molecular dynamics simulation of super-large systems based on physical-model computing, machine learning and high-performance GPU parallel cluster, which won ACM Gordon Bell Prize, the highest award in the field of high-performance-computing application. In their work, IAPCM and UC Berkeley lumped intelligent supercomputing with a physical model for the first time in the world. The use of intelligent supercomputing technology amplified the computing space scale by at least 100 times and improved computing speed by at least 1,000 times over previous work. This signaled progress from the traditional computing model to intelligent supercomputing in scientific computing.

2.2.2 *With Continuous Innovation in Software and Information Technology, Their Industrial Scale Gradually Grows*

In recent years, the software and IT service industry maintained rapid development in China. According to the data of the Ministry of Industry and Information

Technology,[1] in 2020, the revenue scale of the software and IT service industry in China reached 8,161.6 billion *yuan*, with a year-on-year increase of 13.3%. Relevant domestic enterprises remain committed to the fields of software and IT, with a substantial number of achievements put into use.

OS products come into use successively. In desktop and server OS, some products are staged, whose compatibility and adaptability are continuously improved. Deepin optimizes Deepin-wine and develops patches for some application software. With an enhanced Linux kernel, NeoKylin is widely used in energy, finance, transportation and other industries. In mobile OS, many domestic enterprises launch products with differentiated advantages in various segmented fields.

Industrial software advances steadily. In February 2021, industrial software was listed in the Ministry of Science and Technology National Key R & D Program Key Projects. China focuses on supporting four fields, i.e., manufacturing digital ecology & basic cutting-edge technologies, product lifecycle core software, intelligent factory technology & system and industrial-collaboration technology platform. With a complete range of products, industrial software product owns great development potential in the high-end field. Leading enterprises in the manufacturing-software industry made considerable progress, with overall market scale rising. In 2020, the scale of the domestic industrial software market realized around 210 billion *yuan* and would maintain a growth rate of *c.* 16% in the next few years, as predicted.

Cloud integration means a development direction in the future. Industrial software tends to be transferred to the cloud. In the model of deployment, industrial software turns from the enterprise's internal business to the private cloud, public cloud and hybrid cloud. In software architecture, industrial software turns from tight coupling to loose coupling, towards component-based, platform-based and service-based development. In the operating platform, industrial software turns from a PC business to a multiple-mobile-OS-based business. In the development of domestic industrial software, China needs to give play to the pulling effect of enterprise-level industrial applications and technological advantages of industrial software enterprises, and establish a good industrial software ecosystem.

2.2.3 Open Source Ecology Accelerates the Development of Software and Hardware

Recently, China's open-source technology developed swiftly and Open Source Ecology formed gradually. In the open source field, China moved from the stage of learning and use to the stage of participation and innovation, and constantly accelerated the development of domestic software and hardware. Based on Open Source Ecology, numerous enterprises partake in international cooperation and play a part in major projects (rather than minor projects) after sharing open-source information.

[1] Data Source: https://www.miit.gov.cn/gxsj/tjfx/rjy/art/2021/art_4784b23af7054a9eb70004da21352b82.html.

More open-source projects initiated by Chinese developers have come to the international stage. Some Chinese IT companies and Internet enterprises that embrace open-source technology and have great strength develop into major contributors to global open-source projects. Open-source hardware possesses strong extensibility and great convenience, with the unique advantage of technological innovation. Centering on cutting-edge research based on open-source instruction-set architecture RISC-V, domestic enterprises steadily promote the technological innovation of open source hardware. Relevant applications cover mobile terminal, industrial Internet, smart home appliances, communication and storage devices, and the applications in the vertical fields and specific use cases continue to deepen.

2.3 Cutting-Edge Technologies

In 2020, the performance of AI-framework software further improved. Cloud computing displayed a development trend of hybrid cloud, edge cloud and container cloud. With cloud native key technologies widely applied, industrial Internet technology matured. Domestic researchers attained remarkable achievements in the research and exploration of quantum communication, quantum computing and quantum measurement.

2.3.1 AI Continues to Develop Iteratively and New Exploration Continues to Advance

AI technology continues to develop iteratively, and innovative achievements in relevant framework software and chip continue to emerge. Over the past year, China's AI framework software kept improving in performance, professionalization and application. As digital technology continues to advance in China, the development of domestic AI chips features diverse business models and forms.

In the AI field, relevant research institutions have conducted studies and exploration and produced significant achievements, such as the Neural Network High-Speed Training System of Peking University and the Brain-Inspired Computing System of Tsinghua University. In December 2020, the research team from Peking University realized a neural network high-speed training system based on the randomness of phase change memory (PCM), which effectively addressed the problems of huge time & power consumption in the training process of the artificial neural network as well as the use in System-on-Chip (SoC). PKU's training system performs well in the training of large convolutional neural networks . In October 2020, the research team from Tsinghua University firstly proposed the idea of "neuromorphic completeness" as well as a system hierarchy for brain-inspired computing with software and hardware decoupled. Via theoretical demonstration and prototype experiment, the TSU research team proved the completeness and compilation feasibility of the hardware of the system, expanded the application scope of the brain-inspired computing system, and enabled it to support general computing. The research results were published in *Nature* in October 2020.

2.3.2 *Cloud Computing Technology Advances Steadily, with High-Level Projects Continuously Launched*

For nearly a year, as the digital transformation of various industries accelerated in China, the new-generation IT industry represented by cloud computing faced great opportunities for leaps-and-bounds development.

1. Cloud Computing Generally Presents a Development Trend of Hybrid Cloud, Edge Cloud and Container Cloud

Hybrid cloud is already in the pipeline. No matter for enterprise-level IT manufacturers or for pure public cloud manufacturers, hybrid cloud becomes the primary consideration in facilitating corporate user's choice of cloud. Edge cloud is coming up with new breakthroughs. Manufacturers from public cloud, AI, IT, system integration and other fields actualize edge cloud solutions in a faster and deeper way. Cloud-edge collaboration continues to advance, which furthers the integration of AI with cloud, IoT and 5G. Container cloud kindles cloud innovation. Since public cloud manufacturers enthusiastically accomplish the cloudification of container technology, the idea of container cloud has aroused widespread attention from the industrial circle. DevOps integration becomes more intelligent, which popularizes container technologies like Docker and K8S and advances the innovation on PaaS-layer cloud delivery model.

2. Cloud Native Key Technologies are Extensively Adopted

Cloud native technologies serve as an important engine to drive business growth. As new infrastructure that backs digital transformation, cloud native displays unique advantages in emerging fields like AI, big data, edge computing and 5G. As CAICT's survey report suggests, cloud native key technologies represented by container, microservice, service-mesh and DevOps are widely adopted. Enterprises that use cloud native key technologies can adjust their business models in a deeper way, which strengthens corporate scalability and extensibility of enterprises. Cloud native technologies are applicable to large & medium-sized enterprises with the data flow. Enterprises directly cloudify their business architecture, which on the one hand, remarkably reduces their costs, and on the other hand, helps enterprises to better offset their weaknesses and enlarge their business.

3. Kubernetes and Container Become Mainstream Technological Roadmap of Cloud Computing in China

Kubernetes will develop into the core of all technological infrastructure software and the unified tool for the entire infrastructure management platform. Presently, the cloud native infrastructure goes beyond container management. For example, higher requirements are raised for storage. Storage manufacturers integrate their storage solutions with Kubernetes via CSI. Some new container native storage solutions have emerged in communities. Distributed storage systems run directly on Kubernetes that manages and arranges storage systems. Additionally, some container native virtualization technologies come to rise, which directly use Kubernetes to manage and arrange virtual machines and integrate applications that run in virtual machines and containers on one platform for management.

2.3.3 The Construction, Application and Innovation of Industrial Internet Infrastructure Improve Side by Side

In recent years, the construction of industrial Internet infrastructure advanced firmly in China, with the application & innovation ecology continually growing. In December 2020, the Ministry of Industry and Information Technology successively issued the *Industrial Internet Innovation and Development Action Plan (2021–2023)* and *Measures for the Administration of Industrial Internet Identification* to further improve relevant policy support and facilitate the rapid development of industrial Internet in China.

1. The Development of Industrial Internet Breeds Technological Innovation

In recent years, the industrial Internet developed from single-field intelligence to entire-field intelligence. In particular, in manufacturing industries with favorable information foundations, such as automobile, consumer electronics, brand clothing, steel, cement and chemical industry, entire-field-intelligence applications that connect the supply chain, production, assets, logistics and sales will emerge on a large scale. In October 2020, the China Academy of Industrial Internet released *The Top Ten Most Growing Technologies of Global Industrial Internet in 2020–2021*, which included low-code development technology of industrial systems and ultra-intensive heterogeneous technology oriented to industrial fields. Digitization, networking and intellectualization are core elements of the industrial Internet . With prominent advantages in cloud computing, big data, AI and other underlying technologies, industrial giants dominate industrial Internet platforms and maintain their leading role in industrial digital upgrading.

2. Key Technologies Develop and Mature

With technological upgrading and iteration of machine vision, machine learning, language recognition, traceability & anti-counterfeiting and edge computing, the integration and innovation of the industrial Internet will be further upgraded. Industrial Internet solution spans multiple disciplines and technologies and integrates industry, software & hardware facilities and network connection, becoming the cornerstone for the digital transformation and upgrading of the industrial field. Presently, the conditions for software & hardware facilities are basically created, and the development of industrial software is in full swing. Particularly, the supply capacities of software and solutions for specific industries are improved, edge-information facilities like new-type intelligent device, intelligent connection modules and embedded software are developed, and data integration, protocol interface of network connection, data transfer & filtering, as well as edge-information-processing capacity with high interaction and high real-time requirements, are continuously strengthened.

3. The Ecology of the Industrial Internet Is Taking Shape

Under the guidance of the Ministry of Industry and Information Technology, in 2016, more than 100 organizations in the areas of industry, information & communication industry and Internet jointly initiated and established the Alliance of Industrial Internet. The tremendous market of the industrial Internet attracts all parties in the industrial chain. Joint innovation center and joint open laboratory signal the innovation of the cooperation model, provide industrial Internet solutions for industrial enterprises and promote the digital, networked and intelligent transformation of industrial enterprises.

2.3.4 The Exploration of Cutting-Edge Quantum Information Technology is Forwarded Stably

In recent years, China has made progress in the research and exploration of quantum communication, quantum computing and quantum measurement. Regarding the application and exploration of quantum information networks, Chinese researchers have confirmed results in theoretical exploration and conceptual experiments. The research and application of quantum communication highlight quantum cryptography networks, and reap the first fruits of research and experiment. In quantum computing, in May 2021, the 62-bit programmable superconducting quantum-computing prototype "Tsu Ch'ung-chih" was successfully developed, and on this basis, the programmable two-dimensional quantum walk was realized. Relevant research findings were published in *Science*, an international academic journal.

2.4 New Technologies and New Applications

In 2020, new applications of new-generation IT continued to emerge. New applications in the fields of automatic driving and vaccine R & D are expected to alter human living habits and lifestyles.

In automatic driving, China attaches importance to the development of automatic driving and regards it as a prerequisite for the upgrading of the automobile industry. In the past year, China has issued relevant incentive policies to regulate the healthy development of the automatic driving industry with national policies. Stimulated by a series of relevant policies, Internet-technology enterprises, traditional automobile enterprises and technology-based start-ups make a sortie into the field of automatic driving founded on their advantages in capital, technology, channel and other resources.

New technologies accelerate the R & D of vaccines and drugs for COVID-19. Since the COVID-19 pandemic rages globally, the vaccine remains a hard nut to crack for various countries. However, the development of vaccines and drugs requires a longer cycle and more investment. Traditional large-scale experiments prove time-consuming and costly, with an inadequate supply of original products every year. As technologies like supercomputing, AI and cloud computing continue to iterate, their technological application may improve the efficiency of such links as screening compounds, establishing disease models, setting new targets, discovering lead compounds and optimizing lead drugs, reducing repetitive labour and time consumption in clinical research, raise the R & D efficiency, and foster the popularization of medical services and drugs. In 2020, many enterprises used new technologies like AI to accelerate the R & D of the COVID-19 vaccines.

Chapter 3
The Development of Digital Economy

3.1 Outline

Under the dual impact of COVID-19 and the global economic downturn, China's digital economy is growing against headwinds. In 2020, the scale of the digital economy in China totaled 39.2 trillion *yuan*, with a year-on-year increase of 9.7%, accounting for 38.6% of China's GDP.[1] In the "13th Five-Year Plan", the CAGR of the scale of China's digital economy reached 16.08%, whose proportion in China's GDP rose by 11.1%. Digital economy powerfully and effectively supports the prevention & control of the COVID-19 epidemic as well as socio-economic development, and becomes a key driving force for stable economic growth.

In the past year, digital industrialization continued to progress. In 2020, the scale of digital industrialization in China achieved 7.5 trillion *yuan*, with an increase of 0.4 trillion *yuan* over 2019, occupying 19.1% of the scale of the digital economy. Industrial digitization keeps deepening. In 2020, the scale of industrial digitization reached 31.7 trillion *yuan*, with an increase of around 3 trillion *yuan* over 2019, accounting for 80.9% of the scale of the digital economy. The digital penetration and transformation of the primary, secondary and tertiary industries continue to accelerate, with doubly-fast industrial development. The penetration rate of the digital economy into agriculture, industry and service industries realized 8.9, 21 and 40.7% respectively, up by 0.7, 1.6 and 2.9% year on year.

2021 starts the "14th Five-Year Plan". The *Outline of the 14th Five-Year Plan for National Economic and Social Development* proposes to forge new advantages of the digital economy and strengthen new engine of economic development. In the "14th Five-Year Plan", China will further augment the innovation and application of key digital technologies and buttress the reform of production mode with digital transformation.

---

[1] China Academy of Information and Communications Technology (CAICT): *White Paper on the Development of China's Digital Economy, April 2021*, http://www.caict.ac.cn/kxyj/qwfb/bps/202104/P020210424737615413306.pdf.

Chinese Academy of Cyberspace Studies, *China Internet Development Report 2021*,
https://doi.org/10.1007/978-981-19-9311-4_3

3.2 Macro Environment for Digital Economy is Continuously Optimized

The CPC Central Committee with comrade Xi Jinping at its core, coordinates & plans the overall situation of economic and social development, and clarifies the goals and tasks of digital economic construction and development in the "14th Five-Year Plan", which provide a fundamental guideline for the development of the digital economy.

3.2.1 Macro Strategy Continuously Improves

China attaches great importance to the strategic role and significance of the digital economy. In September 2020, the General Office of the State Council issued the *Opinions on Accelerating the Development of New Types of Consumption Driven by New Business Forms and Patterns*, which proposed to cultivate a number of new types of consumption demonstration cities and leading enterprises and significantly increase the proportion of online retail sales of physical goods in total retail sales of social consumer goods by 2025. New consumption forms and models like "Internet + service" have become popular and mature. In November 2020, the General Office of the State Council released the *Implementation Opinions on Promoting the Innovative Development of Foreign Trade*, proposing to promote the digital and intelligent development of the supply chain, actively advance the construction of cross-border e-commerce comprehensive pilot area, quicken the development of new business forms like cross-border e-commerce, accelerate the development of trade digitization, build digital public service platform for trade, and serve the digital transformation of enterprises. In March 2021, The *Outline of the 14th Five-Year Plan for National Economic and Social Development* stated that China would expedite the construction of the digital economy, digital society and digital government, boost the deep integration of digital technologies and real economy, enhance the innovation and application of key digital technologies, advance digital industrialization and foster the digital transformation of industries. In May 2021, the National Development and Reform Commission and other departments jointly published the *National Integrated Big Data Center Collaborative Innovation System Computing Hub Implementation Plan* to facilitate the green and high-quality development of data centers and build a national computing hub system. In June 2021, the National Bureau of Statistics published the *Statistical Classification of Digital Economy and Its Core Industries 2021*, which clarified the basic scope of China's digital economy and divided it into five categories, i.e., digital product manufacturing industry, digital product service industry, digital technology application industry, digital factor-driven industry and digital efficiency-improvement industry. This laid a standard foundation for the statistical needs of the digital economy among the CPC Central Committee, the State Council, CPC Committees and governments at all levels and the society.

3.2.2 Local Policies Are Continually Implemented

Under the guidance of national policies, local governments continue to strengthen the strategic planning of the digital economy and determine development goals and paths of the digital economy. In 2020, national and provincial policy systems basically matured. In China, 29 provinces (autonomous regions and municipalities directly under the Central Government) formulated special policies on the digital economy, including more than 60 digital economy development action plans, industrial planning and subsidy policy. Beijing released the *Beijing Action Guideline for Promoting the Innovation and Development of Digital Economy 2020–2022*, which proposed to increase the proportion of the added value of digital economy in the regional GDP of Beijing to 55% and build Beijing into a pilot area and demonstration zone for the development of the national digital economy by 2022. Sichuan published the *Work Plan of Sichuan on Building the National Digital Economy Innovation and Development Pilot Zone*, which stressed that by 2022, the scale of Sichuan's digital economy would exceed two trillion *yuan*, accounting for 40% of the regional GDP. Zhejiang released the *Work Plan of Zhejiang on Building the National Digital Economy Innovation and Development Pilot Zone*, which highlighted that by 2022, the added value of Zhejiang's digital economy would realize more than 4 trillion *yuan*, accounting for more than 55% of regional GDP. Table 3.1 shows relevant policy documents on the digital economy issued by some provinces (municipalities) since the second half of 2020.

Table 3.1 Relevant policy documents on digital economy issued by some provinces (Municipalities) since the second half of 2020

Provinces (Municipalities)	Issuing time	Policy documents
Yunnan	August 2020	Some policies on supporting the development of digital economy in kunming (For trial implementation)
Beijing	September 2020	Beijing action guideline for promoting the innovation and development of digital economy 2020–2022
Hebei	October 2020	Some policies on supporting the accelerated development of the digital economy
Guangdong	November 2020	Working plan for building national digital economy innovation and development pilot zone in guangdong
Jiangsu	November 2020	Opinions on further promoting the development of the digital economy
Zhejiang	November 2020	Work plan of zhejiang on building the national Digital economy innovation and development pilot zone
Shanghai	January 2021	Opinions on comprehensively promoting urban digital transformation of shanghai
Sichuan	January 2021	Work plan of sichuan on building the national digital economy innovation and development pilot zone

3.2.3 Digital Transformation Continues to Advance

In agriculture, in July 2020, the Office of the Central Cyberspace Affairs Commission and six other departments jointly issued the *Notice on Carrying out National Digital Village Pilot Work*, which deployed national digital village pilot work, and proposed to actively explore new models of digital village development, accelerate the construction of agricultural and rural modernization, and promote the comprehensive upgrading of agriculture, the comprehensive progress of rural areas and the comprehensive development of rural residents. In January 2021, the CPC Central Committee and the State Council issued the *Opinions on Comprehensively Promoting Rural Revitalization and Accelerating Agricultural and Rural Modernization*, proposing to implement the digital rural construction & development program, promote the simultaneous planning & construction of rural gigabit optical network, 5G mobile communication, mobile IoT and cities, strengthen the digital and intelligent construction of rural public services and social governance, and develop rural digital inclusive finance.

In industry, in December 2020, the Ministry of Industry and Information Technology issued the *Administrative Measures for Industrial Internet Identification*, which clarified relevant definitions of industrial Internet identification as well as the regulatory responsibilities of relevant departments, optimized digital rules, promoted the orderly construction of industrial Internet identification and resolution system, and invigorated the innovation and development of industrial Internet identification. In February 2021, the Ministry of Industry and Information Technology released the *Industrial Internet Innovation and Development Action Plan (2021–2023)*, which proposed that by 2023, the construction of new-type industrial Internet infrastructure would advance in both quantity and quality, new models and new business forms would be widely applied, and comprehensive industrial strength would significantly improve. In February 2021, the Ministry of Transport, Ministry of Industry and Information Technology and Standardization Administration jointly issued the *Guidelines for the Construction of the National Standards System for the IoV Industry (Related to Intelligent Transportation)*. It proposed that by the end of 2022, more than 20 urgently-needed standards for intelligent transportation in the fields of intelligent transportation infrastructure and traffic information assistance would be formulated or revised, and that by 2025, more than 20 key standards for intelligent transportation in the fields of intelligent management & services and vehicle–road coordination would be formulated or revised, and a standard system that can support IoV application and meet the needs of transportation management and service would be formed.

In the service industry, in August 2020, the Ministry of Commerce issued the *Overall Plan for Comprehensively Deepening the Innovation and Development Pilot of Service Trade*, proposing to explore flexible employment systems and policies compatible with the development of the digital economy and digital trade, facilitate digital business environment, comprehensively explore innovation & development models, expand new business forms and new models, vigorously develop digital trade and improve policies on digital trade. In November 2020, the Ministry of Culture and Tourism, the National Development and Reform Commission and eight other departments jointly published the *Opinions on Deepening the "Internet + Tourism" to Promote the High-Quality Development of Tourism*, proposing to build a number of smart tourism scenic spots, resorts, villages, towns and cities by 2022 and to basically achieve intelligent transformation and upgrading of national 4A level and above tourist attractions and provincial-level and above tourist resorts by 2025, with the total number of tourist reception and the scale of tourism consumption remarkably increased, and the attraction and influence of overseas tourists significantly enhanced. In February 2021, the CPC Central Committee and the State Council issued *the National Comprehensive Three-dimensional Transportation Network Planning Outline*, which proposed to accelerate the improvement of the scientific & technological innovation capacity of transport, promote the digitization and networking of transport infrastructure, and improve the level of intelligent development of transport, and to build modern and high-quality national comprehensive three-dimensional transportation network in an all-round way by the middle of the twenty-first century, with new technologies widely used to realize the digital, networked, intelligent and green development of transport.

3.2.4 The Governance of Digital Platforms is Continuously Enhanced

In 2020, the Central Conference on Economic Work resolved to strengthen antitrust governance and prevent the disorderly expansion of capital. As stated, China "supports the innovation and development of platform enterprises and enhances their international competitiveness", "improves the legal norms on the identification of platform enterprise's monopoly, the collection, use and management of data and the protection of consumer's rights & interests" and "strengthens regulation, optimizes regulatory capacity, and resolutely combats monopoly and unfair competition". In January 2021, the General Office of the CPC Central Committee and the General Office of the State Council jointly issued the *Action Plan for Building a High-Standard Market System*, which clearly proposed to strengthen the regulations on antitrust governance and anti-unfair competition in new business forms like platform economy and sharing economy. In January 2021, the People's Bank of China released the *Regulations on Non-bank Payment Institutions (Draft for Comments)* to strengthen antitrust regulatory measures in the payment field, clarify relevant market scope and the criteria for determining market dominance, and safeguard the market order of fair competition. In February 2021, the Antitrust Commission of the State Council formulated and released *Antitrust Guidelines for Platform Economy*. It provides scientific, effective and targeted institutional rules for strengthening the antitrust regulation on platform economy, treats all market entities equally, fairly and justly, prevents and curbs monopolistic acts in the platform economy, and promotes the standardized, orderly, innovative and healthy development of platform economy, emphasizing that the *Anti-monopoly Law of the People's Republic of China* and its supporting regulations are applicable to all industries.

3.3 Data as a Production Factor Becomes Increasingly Prominent

In March 2020, the CPC Central Committee and the State Council issued the *Opinions on Improving the Systems and Mechanisms for Market-based Allocation of Factors of Production*, which clearly proposed to accelerate the cultivation of the data-factor market and specified the path for further giving play to the role of data as a production factor. To release the value of data as a production factor plays a crucial role in realizing the high-quality development of the Chinese economy.

3.3.1 The Total Data Output Expands Rapidly

In April 2021, the China Academy of Information and Communications Technology (CAICT) and the Chinese Academy of Cyberspace Studies (CACS) co-published the *National Data Resources Survey Report 2020*. As it suggests, in 2019, the total scale of China's data output reached 3.9 ZB, with a year-on-year increase of 29.3%, accounting for 9.3% of the total global data output (42 ZB). The per capita data output in China realized 3 TB, with a year-on-year increase of 25%. In the regional distribution of data output, in 2019, top ten provinces (municipalities directly under the Central Government) were Guangdong, Beijing, Zhejiang, Jiangsu, Shanghai, Shandong, Sichuan, Henan, Hebei and Hunan. Specifically, the data outputs of Guangdong and Beijing were more than three times of national average level (127.4 EB). In per capita data output, top ten provinces (autonomous regions and municipalities directly under the Central Government) were Beijing, Shanghai, Zhejiang, Tianjin, Guangdong, Inner Mongolia, Tibet, Hainan, Jiangsu and Liaoning. In data source & structure, data resources were mainly generated by various equipment held by industrial institutions and individuals. In 2019, the data output of Chinese industrial institutions reached 3ZB, accounting for 76.9% of the total data output of China, and the output of personal data accounted for 23.1% of the total data output of China. In the past three years, the proportion of data output of industrial institutions increased yearly. Data, as a production factor, functions as basic resources for socio-economic development and an engine for innovation. *Vis-à-vis* traditional production factors, data play a more flexible and diverse role and make greater contributions to China's economic growth in the future, as expected.

3.3.2 The Scale of the Data Industry Continues to Grow

The data industry accelerates the marketization of data elements, continuously activates the potential value of data elements, and steps up deep integration with the real economy. In recent years, the big data industry in China continued to grow rapidly and become an important force in promoting high-quality economic development. According to the *Digital China Development Report 2020* released by the Cyberspace Administration of China, the scale of China's big data industry increased from 0.34 trillion *yuan* in 2016 to more than 1 trillion *yuan* in 2020. According to the *Report on the Development of China's Data Factor Market 2020–2021* released by the National Industrial Information Security Development Research Center, the "13th Five-Year Plan" signaled a stage in which the big data industry in China thrived. By the end of 2020, the scale of China's big data industry reached more than one trillion *yuan*, with the big data industry deeply integrated with the real economy and growing industrial development.

3.3.3 Data Element Market Develops Steadfastly

According to the *Digital China Development Report 2020*, the construction of China's data factor market accelerates. In August 2020, the Beibu Gulf Big Data Exchange was established. By the end of 2020, the transaction scale exceeded 15 million *yuan*, with more than 120 registered enterprises and more than 120 million times of data service calls. As the *Report on the Development of China's Data Factor Market 2020–2021*[2] evidence, currently, the data element market that centers on the links such as data collection, data storage, data processing and data circulation grows fast. In 2020, the scale of China's data factor market reached 54.5 billion *yuan*. In the "13th Five-Year Plan", the CAGR of the market scale exceeded 30%. By the end of the "14th Five-Year Plan", the scale of China's data factor market will exceed 174.9 billion *yuan*. By means of technological innovation and coordinated development, China's data factor market has initially formed a standardized and orderly industrial ecology. In terms of technological integration, new technologies represented by federal learning, secure multi-party computing and data sandbox continually spur the development of China's data factor market. With regard to regional development, the pattern of regional division & cooperation in the data element market takes shape. Beijing, Shanghai, Guangzhou, Shenzhen and other cities rely on their advantages in talents and technologies and step up the development of high-end businesses like data circulation and data-technology R & D. Developing areas engage in traditional data services like data labeling and data cleaning by taking advantage of labor-intensive industries.

[2] Data Source: National Industrial Information Security Development Research Center.

3.4 Digital Industrialization Continues to Advance

In 2021, the downward trend of electronic information manufacturing was reversed, with industrial scale increasing substantially. Software and IT service industries bailed out the negative impact of COVID-19, and industrial scale and efficiency comprehensively improved. The overall development of the Internet and related service industries remained stable, the ecology of the Internet industry constantly advanced, and the collaborative transformation of digital and green development continually deepened.

3.4.1 The Scale of Electronic Information Manufacturing Increases Significantly

In 2021, electronic information manufacturing recovered, and relevant indicators improved. In the first half of 2021, the added value of above-designated-scale electronic information manufacturing increased by 19.8% year on year, with a growth rate of 14.1% over 2020. The cumulative export delivery value increased by 17.6% year on year, up by 13.6% over 2020. Investment in fixed assets rose by 28.3% year on year, up by 18.9% over 2020, 12.1% higher than industrial investment. Figure 3.1 shows the growth rates of the added value and export delivery value of the electronic information manufacturing industry from June 2020 to June 2021.

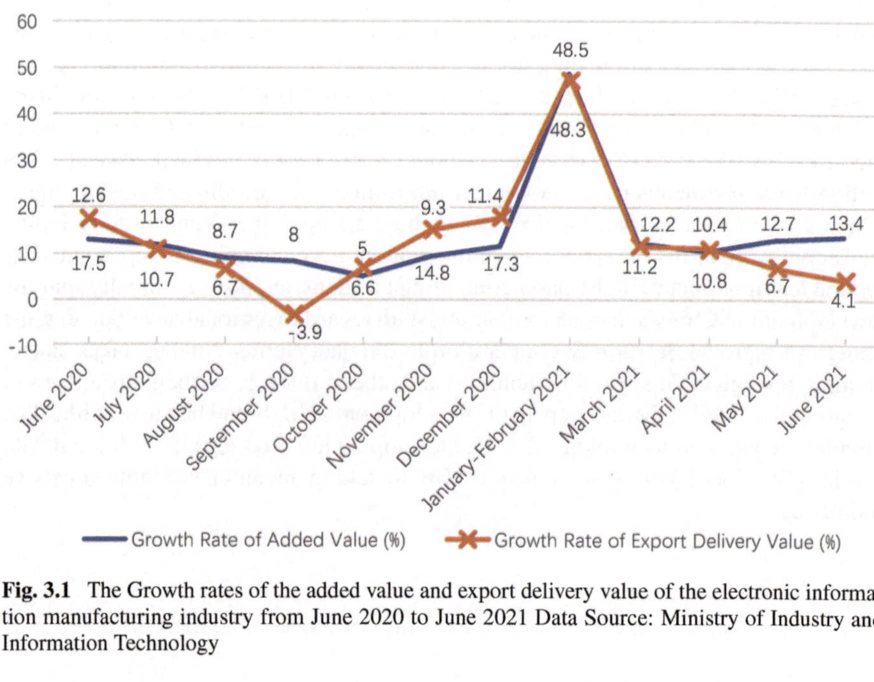

Fig. 3.1 The Growth rates of the added value and export delivery value of the electronic information manufacturing industry from June 2020 to June 2021 Data Source: Ministry of Industry and Information Technology

In June 2021, in relation to the main products of the electronic information manufacturing industry, the output of mobile phones reached 140 million, with a year-on-year increase of 2.9%. To be specific, the output of smart phones reached 110 million, with a year-on-year decrease of 0.6%. The output of microcomputer equipment reached 41.078 million, with a year-on-year increase of 30.9%. The output of integrated circuits reached 30.82 billion, with a year-on-year increase of 43.9%.

3.4.2 *The Scale and Efficiency of Software and IT Service Industries Gradually Improve*

In 2020, software and IT service industries comprehensively raised industrial scale and efficiency against the headwinds of COVID-19. The revenue and profit of software business maintained a rapid growth rate, and the average annual growth rate of revenue outstripped the expectation of the software industry in the "13th Five-Year Plan". Yet, software export remained in the doldrums. More than 40,000 above-designated-scale enterprises engage in software and IT service industries, whose cumulative revenue in software business reached 8,161.6 billion *yuan*, with a year-on-year increase of 13.3%. The total profit of the whole industry realized 1,067.6 billion *yuan*, with a year-on-year increase of 7.8%. The total export value achieved 47.87 billion U. S. dollars, with a year-on-year decrease of 2.4%. The number of employees reached 7.047 million, with an increase of 210,000 over the end of 2019, with a year-on-year increase of 3.1%. The wages of employees totaled 994.1 billion *yuan*, with a year-on-year increase of 6.7%, lower than the average growth rate in 2019.

In the first half of 2021, the revenue of software and IT service industries in China displayed a trend of fast growth. The revenue of the software business reached 4,419.8 billion *yuan*, with a year-on-year increase of 23.2% and a CAGR of 14.7% in the past two years. Specifically, the revenue from software products reached 1,145.1 billion *yuan*, with a year-on-year increase of 20.1%, accounting for 25.9% of the industrial revenue. The revenue from IT service realized 2,831.9 billion *yuan*, with a year-on-year increase of 26%, accounting for 64.1% of the industrial revenue. The revenue from information-security products and services totaled 73.2 billion *yuan*, with a year-on-year increase of 26% (indicating more importance attached to information security). The revenue of embedded-system software achieved 393.6 billion *yuan*, with a year-on-year increase of 12.8%. Figure 3.2 shows the proportions of revenues of software and IT service industries in the first half of 2021.

In the first half of 2021, software and IT service industries in China realized a total profit of 499.9 billion *yuan*, with a year-on-year increase of 13.6% and a CAGR of 7.3% in the past two years. Software exports grew steadily and achieved a total export value of 24 billion U. S. dollars, with a year-on-year increase of 12.2% and a CAGR of 5.7% in the past two years. Specifically, the export volume of outsourcing services reached 6.86 billion U. S. dollars, with a year-on-year increase of 19.8%. The export volume of embedded-system software realized 9.38 billion U. S. dollars, with a year-on-year increase of 4.8%. Figure 3.3 shows the income, total profit and export growth of software and IT service industries in 2020 and the first half of 2021.

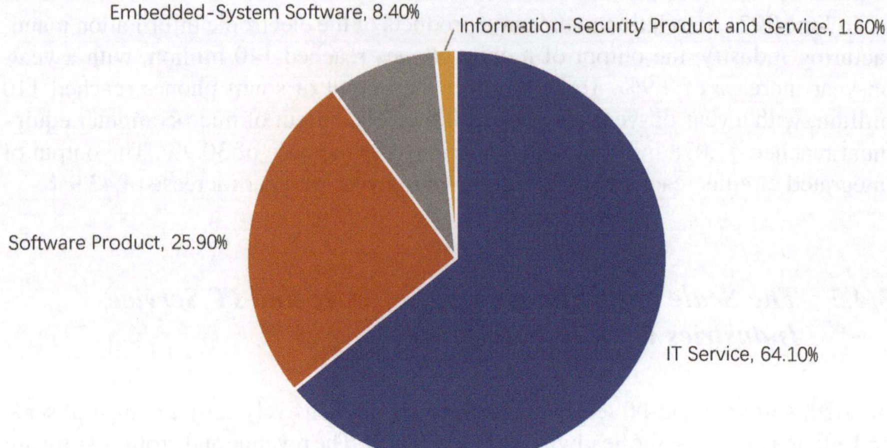

Fig. 3.2 The proportions of revenues of software and IT service industries in the first half of 2021 Data Source: Ministry of Industry and Information Technology

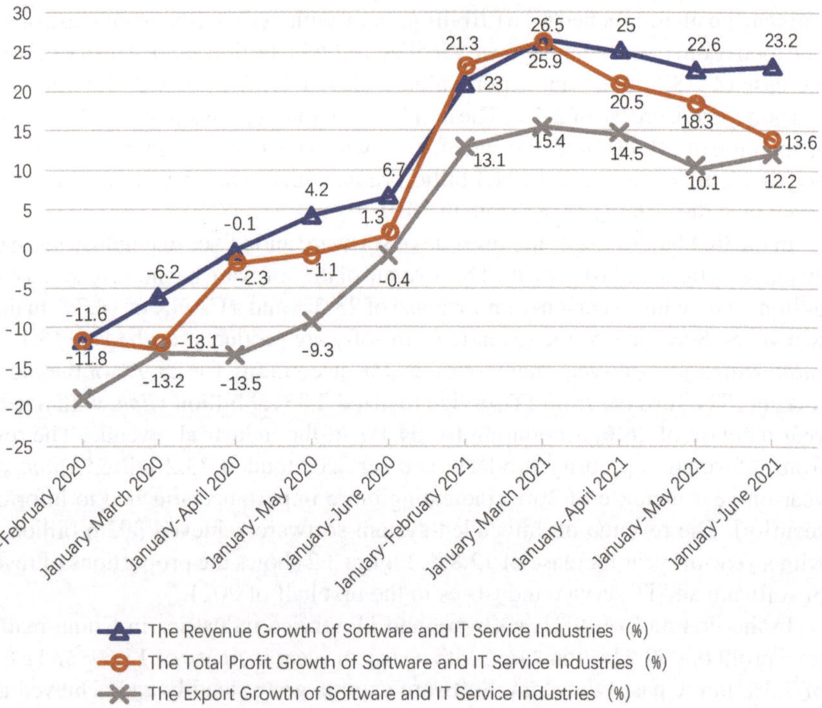

Fig. 3.3 The income, total profit and export growth of software and IT service industries in 2020 and the first half of 2021 Data Source: Ministry of Industry and Information Technology

3.4.3 *Internet Information Content Service Industry Develops Stably*

In 2020, the overall development of Internet and related service industries in China presented a stable trend, and the industrial revenue remained stable yet dipped slightly. According to the data released by the Ministry of Industry and Information Technology, in 2020, the above-designated-scale Internet enterprises and enterprises engaged in related services achieved a business revenue of 1,283.8 billion *yuan*, with a year-on-year increase of 12.5%, lower than that in 2019. The business profit realized 118.7 billion *yuan*, with a year-on-year increase of 13.2%, down by 3.7% over 2019, up by 0.7% in growth rate over 2019. The investment in R & D reached 78.8 billion *yuan*, with a year-on-year increase of 6%, down by 17.1% over 2019. As the quality and development of online entertainment and news-media applications continue to improve, the Internet information content service industry maintains steadfast growth, yet the growth rate slows down. In 2020, the revenue totaled 706.8 billion *yuan*, with a year-on-year increase of 11.5% (down by 11.2% over 2019), accounting for 55.1% of the Internet business revenue. In particular, the growth rate of business revenue in the audio & video service areas dropped, and the high-speed growth trend gradually slowed, with the R & D expenses remaining high. Under the impact of COVID-19, the growth of online games was higher in the early stage and lowered in the late stage. The business income of news & content-service enterprises achieved positive growth. The overall business revenue of the search engine industry declined, and the competition intensified. The business of enterprises engaged in search services stagnated.

In 2021, Internet and related service industries maintained a strong momentum for development, with business revenue growing rapidly. According to the data published by the Ministry of Industry and Information Technology, in 2021, from January to May, the above-designated-scale Internet enterprises and enterprises engaged in related services realized a business revenue of 598.1 billion *yuan*, with a year-on-year increase of 24.4% and a year-on-year growth rate of 9.5%. The business profit reached 60.66 billion *yuan*, with a year-on-year increase of 13.1%. The investment in R & D reached 29.01 billion *yuan*, with a year-on-year increase of 16.3% and a year-on-year growth rate of 6.3%.

Innovation and convergence are strengthened in Internet news media. On the one hand, Internet news media adapt to the development of the times, and continue to innovate the reporting forms, which create a more direct and detailed news experiences to users. On the other hand, Internet news media further strengthen media convergence, enhance user viscosity and improve user experience by building a video communication matrix. By June 2021, the scale of Internet news users in China had reached 760 million, with an increase of 17.12 million over December 2020, accounting for 75.2% of all Internet users. Figure 3.4 shows the scale and proportion of Internet news users in China from June 2018 to June 2021.

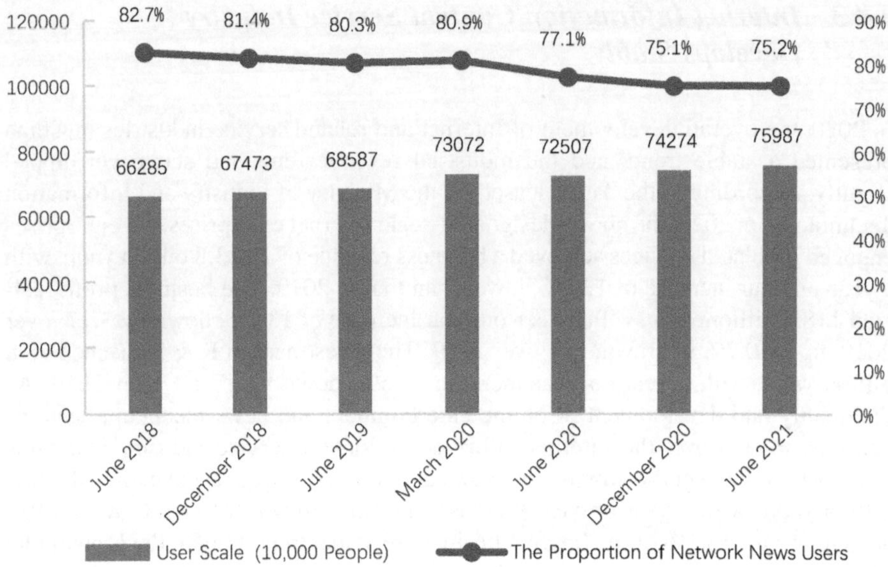

Fig. 3.4 The scale and proportion of internet news users in China from June 2018 to June 2021[3]

Video websites optimize their business models and short-video platforms expand their business layout. Backed by high-quality content, video websites start to optimize their business models. Short-video platforms encourage the production of high-quality content in various ways, enhance user viscosity and make a foray into the comprehensive video business and long-video field, with the scale of short-video user growing rapidly. By June 2021, the number of short video user in China reached 888 million, with an increase of 14.4 million over December 2020, accounting for 87.8% of all Internet users. Figure 3.5 shows the scale and proportion of short video users in China from June 2018 to June 2021.

The online game industry maintains rapid development. By June 2021, the number of online game users in China had reached 509 million, accounting for 50.4% of all Internet users. On the one hand, mobile games in China accelerate the pace of innovation, with the emergence of new popular games. Besides, in terms of independently-developed games, China's international influence and market position keep enlarging and rising. On the other hand, many host games in China achieve cross-field development, with great potential to be tapped in the future. More investment to the R & D in

[3] Data Source: *The 47th Statistical Report on China's Internet Development* released by the China Internet Network Information Center.

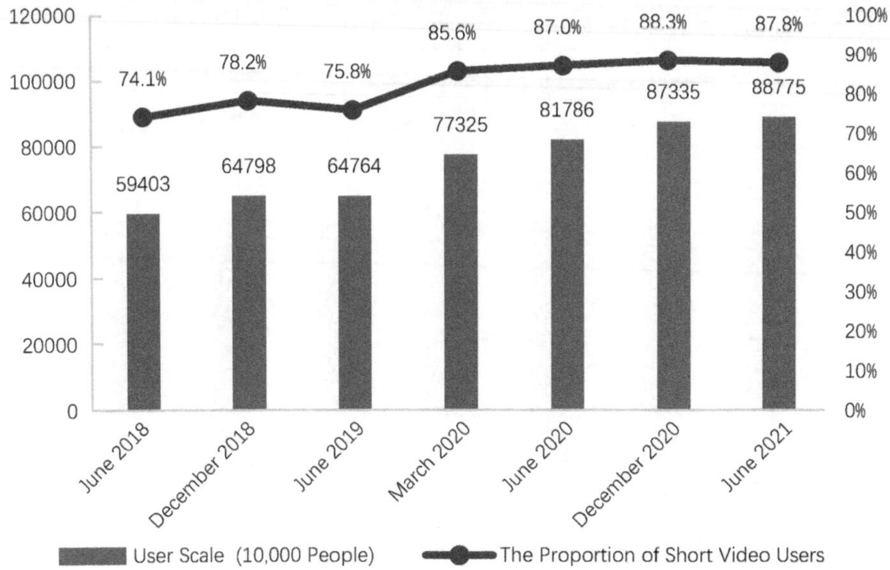

Fig. 3.5 The scale and proportion of short video users in China from June 2018 to June 2021[4]

this regard will promote the development of domestic-related industries and further enhance the international popularity, influence and revenue capacity of China's online game industry. Figure 3.6 shows the scale and proportion of online game users in China from June 2018 to June 2021.

3.4.4 *The Ecology of the Internet Industry is Continuously Diversified and Improved*

The ecology of the Internet industry in China is continuously diversified and improved. New business forms of digital technology continue to emerge. A large number of enterprises in big data, cloud computing and AI achieve innovation-driven development. With a complete industrial production system, the enterprises advance towards global middle & high-end industrial chains. Digital transformation accelerates in all fields, aspects and links of society and economy, and new digital models and business forms mushroom.

After 2021, Internet enterprises saw a new stage of high-quality development. The number of Internet listed enterprises continued to grow, with the market value

[4] Data Source: *The 48th Statistical Report on China's Internet Development* released by the China Internet Network Information Center.

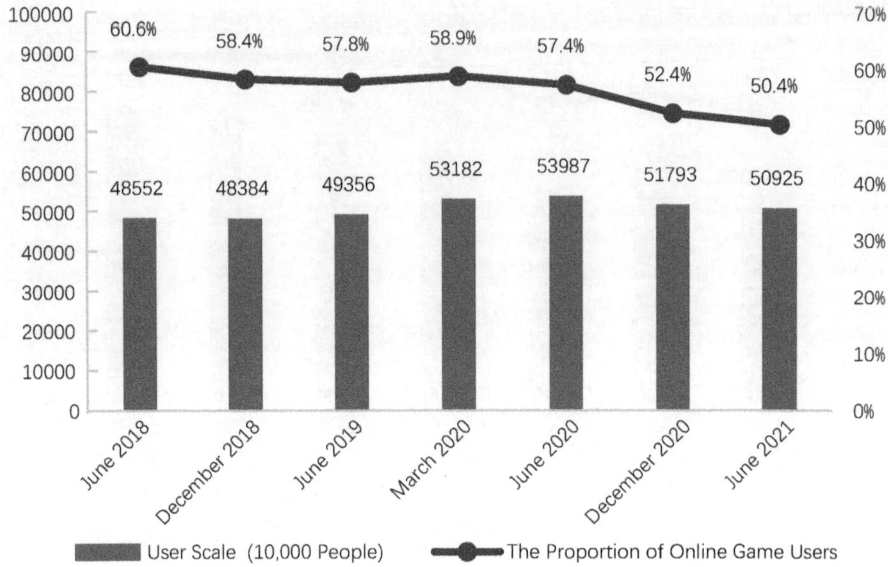

Fig. 3.6 The scale and proportion of online game users in China from June 2018 to June 2021[5]

reaching a new record. The China Academy of Information and Communications Technology (CAICT) published the *Operation of Internet Listed Enterprises in China in the Second Quarter of 2021*. As it evinced, by June 2021, there were 187 Internet listed enterprises in China, with an increase of 5.1% over the end of 2020. The total market value of Internet listed enterprises reached 18.8 trillion *yuan*, with an increase of 3.9% over the end of 2020 and a chain growth rate of 0.2%. Specifically, the market value of top ten Internet enterprises accounted for 77.4% of the total market value, down by 3.5% over the end of 2020. The cluster development of Internet enterprises takes initial shape. Around 80% of Internet listed enterprises and internet-information Unicorns are distributed in Beijing, Shanghai, Guangdong and Zhejiang. Simultaneously, the reform of the registration-based system is deepened and normalized in an all-round way, and the delisting provision is established, which creates new development opportunities for growth-oriented innovation-driven start-ups.

In the first half of 2021, as overall prevention & control of COVID-19 was on the mend, remarkable achievements were made in the socio-economic field. Generally speaking, the investment & financing in the Internet field took a turn for the better over 2020, and the chain growth rate of financing steadily decreased. The China Academy of Information and Communications Technology (CAICT) released the

[5] Data Source: *The 48th Statistical Report on China's Internet Development* released by the China Internet Network Information Center.

Operation of Internet Investment and Financing in the Second Quarter of 2021. As suggested, 554 times of investment & financing were completed in the second quarter of 2021, with a year-on-year increase of 15.7% and a chain growth rate of 9.7%. The investment & financing amounted to 15.56 billion U. S. dollars, with a year-on-year increase of 66.6%; yet, the chain growth rate dropped by 8.1%. In 33 times, large-scale investment & financing amounted to more than 100 million U. S. dollars, with total financing of 11.8 billion U. S. dollars, accounting for 79% of the total financing in the quarter. In specific fields, the investment & financing of e-commerce and travel tourism increased significantly in the second quarter of 2020. As an important platform, the STAR market supports new-generation high-tech industries and strategic emerging industries, and provides financing channels for many high-quality new-economy and high-tech enterprises. From January to June 2021, the number of listed enterprises on the STAR market reached 86,[6] and 34 enterprises engaged in the Internet information field, accounting for 39.5%.

3.4.5 Coordinated Digital and Green Transformation and Development Continue to Deepen

In September 2020, at the 75th Session of the UN General Assembly, Chinese President Xi Jinping made the solemn promise that before 2030, China's carbon dioxide emissions would peak ("Carbon Peak"), and before 2060, China would achieve carbon neutrality. Digital development and green development complement each other. The former leads the latter, which can effectively improve the utilization level and information-service capacity of data resources in high-energy-consumption industries and accelerate the pace of green development. The latter promotes the former, which drives the digital transformation of key industries. Coordinated digital and green transformation provides a strong momentum that helps China achieve the goals of "Carbon Peak" and "Carbon Neutrality".

To carbon emission reduction or zero emission, Chinese Internet enterprises respond positively and make commitments to boost green and sustainable development. On January 13, 2021, Tencent launched "Carbon Neutrality Program", which used AI to upgrade new ideas like global infrastructure, water-energy connectivity, flying cars, energy Internet and digital water and provided new sustainable development solutions for solving the challenges of food, energy and water. Baidu reduces the total carbon emissions of data centers by using clean low-carbon energy and improving energy efficiency. As relevant data indicated, in 2020, the number of wind power contracted by Baidu increased by 50% over 2019 and reached 45 million kW · h.[7] In 2020, JD proposed the goal of reducing carbon emissions by

[6] Data Source: Data from the IPO Research Center, www.lanfucaijing.com.

[7] Data Source: https://m.k.sohu.com/d/515896480.

50% by 2030. Huawei released digital-energy zero-carbon network solutions (e.g., minimal sites, minimal computer rooms, minimal data centers and ubiquitous green power) to help operators realize the zero-carbon network strategy and accelerate global green and sustainable development.

In 2021, relevant policies and documents were issued swiftly and a national carbon- emission trading-rights market was opened, which provided relevant services on functional coordination, working coordination and mechanism coordination for the realization of "Carbon Peak" and "Carbon Neutrality". In January 2021, the Ministry of Ecology and Environment issued the *Guiding Opinions on Coordinating and Strengthening the Work Related to Climate Change and Ecological Environmental Protection*. In the power grid, coal, transportation and other key industries, actions on "Carbon Peak" and "Carbon Neutrality" are studied and deployed. In the "14th Five-Year Plan", various provinces (autonomous regions and municipalities directly under the Central Government) clearly state their goals and tasks on "Carbon Peak" and "Carbon Neutrality", and quickly forge policies and social environment for achieving "Carbon Peak" and "Carbon Neutrality". In June 2021, the Torch High Technology Industry Development Center of the Ministry of Science and Technology, in collaboration with several national high-tech zones, released the *Declaration of Action on "Carbon Peak" and "Carbon Neutrality" of National High-Tech Zones*, to enhance the support for the R & D of cutting-edge and revolutionary green low-carbon technologies and contribute to the realization of the goals of "Carbon Peak" and "Carbon Neutrality" in China. In June 2021, the first national green technology trading center was officially opened in Hangzhou, which highlighted key technologies in China's low-carbon transformation, guided China's innovation on green technology, and accelerated the transformation and application of achievements. In July 2021, the national carbon-emission trading market was opened. The carbon emissions of enterprises included in the first-batch carbon markets exceeded four billion tons of carbon dioxide. This signifies that once the carbon-emission trading market in China is launched, it will become the largest carbon market that covers greenhouse gas emissions in the world. To adopt a market mechanism to control and reduce greenhouse gas emissions means an important institutional innovation that fosters green and low-carbon transformation of the economic development model.

3.5 Industrial Digitization Continues to Deepen

IT applications are accelerated, the digital industry is further integrated with real economy, and the digital transformation of the economy and society is advanced.

3.5.1 The Level of Agricultural Digitization Continuously Improves

1. The Construction of Agricultural & Rural Big Data Produces Initial Results

The data resource collection system continues to improve, and the level of data resource sharing, business collaboration and data openness continues to rise. The application scope of data resources expands. The data-application system of the whole industrial chain takes initially shape, which centers on the data collection, analysis, release and service of single varieties of key agricultural products such as oil, natural rubber, sugar cane, cotton, apple, soybean and pig. The big data system is applied to more fields, and the national agricultural product quality & safety traceability management information platform is fully promoted. According to the *Report on Digital Development in Rural China 2020*, by July 2020, there were 94,100 production & operation entities on the national agricultural product quality & safety traceability management information platform, with 981 product types and 113,000 data uploaded by enterprises.

2. The Level of Agricultural-Production Digitization Continually Rises

Digital technology boosts rural modernization and development, accurately controls the quality of agricultural products, improves the standardization of agricultural products and comprehensively empowers agricultural segmented areas. The information-based construction of the planting industry attains remarkable achievements, and the digital construction of administrative management platforms sustainably advances. By the first half of 2020, the platform had collected 1,511 pesticide production licenses, 230,000 business licenses, *c.* 4.8 billion production traceability codes, and 157,900 planting users registered to install the agricultural materials purchase, sales and storage system. The livestock-farm direct connection & reporting system actualizes horizontal interconnection and provincial-ministerial interaction. By the first half of 2020, the number of large-scale farms registered on the platform had reached more than 800,000, involving around 12 million pieces of information. The construction of fishery information systems accelerates, and the production & marketing platform becomes more popular. The digital breeding platform is successfully applied, which forges a solid foundation for big data management of the seed industry. The *Specifications on the Data Description of Digital Seed Industry (Draft)* comes to effect, which continuously enhances the standardization of seed industry data. The digitization of agricultural machinery equipment keeps advancing. By September 2020, China had financed 12,000 farmers to purchase 15,000 agricultural BDS terminals, supported 20 provinces in conducting standardized application pilot of plant-protection UAVs, and subsidized 5,000 farmers in purchasing around 10,000 plant-protection UAVs. In China, 45 agricultural machinery enterprises have adopted BDS terminals as standard equipment.[8]

[8] Data Source: *Report on Digital Development in Rural China 2020.*

3. Poverty-Alleviation-Through-Internet Produces Remarkable Results

In poverty alleviation, China utilizes digital technologies like big data and cloud computing, builds big data-based platforms like "poverty alleviation cloud", and accurately collects the causes of poverty and information of poverty-stricken populations so as to further improve the accuracy and transparency of poverty alleviation. The information service system of poverty-alleviation-through-Internet takes basic shape, and the monitoring, early warning and rescue services for the return-to-poverty and new poverty-stricken populations continuously improve with information-based technologies. Digital economy expands new space for rural development. County heads and town heads become livestreaming hosts and sell goods, with livestreaming as "new farm work". As poverty-alleviation-through-Internet continues to deepen, the participation degree of Internet users into poverty-alleviation-through-Internet significantly increases. By December 2020, the proportion of Internet users who participated in poverty-alleviation-through-Internet and purchased characteristic agricultural products in poverty-stricken areas had reached 48.2 and 39.7% respectively, up by 4.3 and 16.7% over March 2020.[9]

4. New Business Forms of the Rural Digital Economy Develop Rapidly

In line with the deployment of the State Council, in 2020, the Ministry of Agriculture and Rural Affairs, working with relevant departments, organized and implemented "Internet +" Agricultural Products in Villages Sold in Cities Program, and accelerated the construction of the supply-chain system, operation-service system and support-guarantee system suitable for the online sales of agricultural products. Digital technologies are fully integrated with e-commerce, livestreaming and other Internet applications, which further shape and enhance brand influence, with tremendous progress in E-commerce in the Rural Areas Comprehensive Demonstration Program. According to the data of the Ministry of Commerce, in 2020, e-commerce came to rural areas, which achieved the full coverage of 832 national poverty-stricken counties and invigorated the rural consumer market. National rural online retail sales increased from 180 billion *yuan* in 2014 to 1.79 trillion *yuan* in 2020. As the construction of the digital village continues to advance, the digital level of public services in rural medical care, education, finance and other fields continues to rise. In particular, telemedicine realizes the full coverage of county-level hospitals in poverty-stricken counties at the national level. Acquiring-Knowledge-Through-Internet quickens. The Internet access rate of primary and secondary schools (including teaching bases) in China increased from 79.37% at the end of 2016 to 100% at the end of 2020. The coverage rate of basic financial services in administrative villages nationwide reached 99.2%.

[9] Data Source: *The 47th Statistical Report on China's Internet Development* released by the China Internet Network Information Center.

3.5.2 *Industrial Digital Transformation Continues to Advance*

Since 2020, China has steadily promoted the construction of industrial Internet infrastructure, constantly improved the level of industrial integration and development, and empowered the digital transformation of key industries. With the acceleration of the digital transformation of industry and manufacturing, the role of digital technology in the intelligent upgrading & transformation of industry and manufacturing and the green transformation and development becomes more prominent. It is an important task to take an intelligent, efficient, clean and intensive green development path for industrial development in China.

1. The Economic Scale of the Industrial Internet Keeps Expanding

By deepening the integration of Internet, big data, AI and real economy, China continues to build an industrial Internet ecosystem and guide industrial transformation & upgrading with scientific & technological innovation. According to the *Industrial Development Report 2020* released by the CAICT, in 2020, the economic added value of China's industrial Internet reached *c.* 3.1 trillion *yuan*, with a year-on-year growth of c. 47.9%, whose contribution rate to GDP growth exceeded 11%.

2. Industrial Internet Empowers the Development of Key Industries

In the fields of coal, steel and machinery, industrial Internet continuously improves production efficiency, reduces operating costs and ensures safe production via technologies like AI empowerment, 5G intelligent manufacturing and business system cloudification. This generates new impetus to industrial development, supports the transformation of the industrial production model, and robustly promotes the digital transformation of industries and enterprises. In terms of platform innovation, the scale of industrial Internet platforms continues to grow, and innovation-oriented solution continues to emerge. By March 2021, there were more than 100 industrial Internet platforms with industrial and regional influence, more than 70 million connected industrial devices, more than 590,000 industrial apps and more than one million cloudified enterprises.[10] The solution that lumps AI with industry becomes a new development trend. In industrial equipment, the innovation of industrial robots plays a leading role in the rapid development of unmanned factories and industrial logistics. In 2020, the output of industrial robots in China reached 237,000, with a year-on-year increase of 19.1%. Welding robots gradually cover aerospace, high-end equipment, rail transit and other fields. Industrial robots can realize full-automatic sorting of category warehouses. With flexible industrial logistics solutions constantly implemented, industrial logistics achieves rapid development. In the field of intelligent construction, in June 2021, the dam of Baihetan Hydropower Station, the largest hydropower station under construction in the world, was poured to the top. In the project, industrial technologies like intelligent water supply and intelligent grouting were applied and integrated with cutting-edge information technologies like DIM,

[10] Data Source: Ministry of Industry and Information Technology.

big data and IoT, to build a high-sensitivity & rapid-response intelligent construction system and construction environment, achieve 24 h dynamic supervision, evaluation and early warning, and ensure the most objective and direct data-aided scientific on-site decision-making.

3.5.3 *The Digitization of the Service Industry Accelerates*

In 2020, economic activities were quickly transferred to online platforms, and the demand for digital applications & services was significantly enlarged. From January to November 2020, the business revenue of the above-designated-scale enterprises in strategic emerging service industries increased by 8.6% year on year, 7% higher than that of the above-designated-scale enterprises in the service industry.[11]

1. Online Education Industry Develops Healthily

By June 2021, online education users in China had reached 325 million, with a decrease of 16.78 million over December 2020, accounting for 32.1% of all Internet users. The online education industry attracts many institutions and capital resources, and market competition intensifies. From January to October 2020, online education enterprises in China increased 82,000, accounting for 17.3% of the total education industry. From January to November 2020, the financing of the online education industry totaled around 38.8 billion *yuan*, with a year-on-year increase of 256.8%. Figure 3.7 shows the scale and proportion of online education users in China from June 2018 to June 2021.

2. The Online Medical Industry Faces Important Development Opportunities

Hammered by the COVID-19 pandemic, online medical services display prominent advantages. New-type medical forms, such as online diagnose-through- interrogation, video diagnose-through-interrogation, online drug purchase and "Internet +" medical insurance payment quickly develop and expand. According to the data released by the National Health Commission, by October 2020, there were more than 900 Internet hospitals in China. More than 5,500 above-second-level hospitals can provide online services. Telemedicine collaboration network covers more than 24,000 medical institutions in all prefecture-level cities. The scale of online medical users continues to enlarge and the penetration rate of users continues to rise. According to *The 48th Statistical Report on China's Internet Development* released by the China Internet Network Information Center (CNNIC), by June 2021, the scale of online medical users in China had reached 239 million, with an increase of 24.53 million over December 2020, accounting for 23.7% of all Internet users. Catalyzed by favorable policies and market demands, the traffic of various medical platforms increased dramatically, and the medical industry came to a new stage of development.

---

[11] Data Source: National Bureau of Statistics.

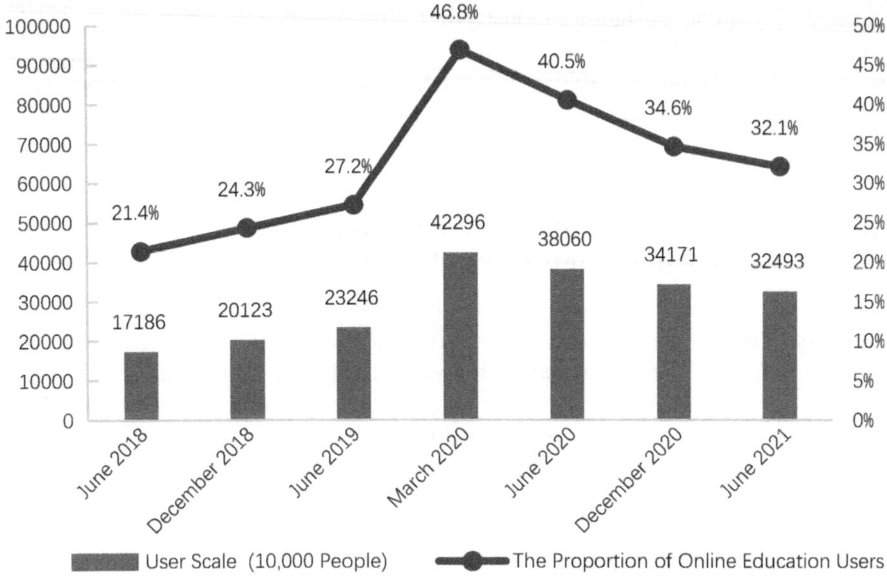

Fig. 3.7 The scale and proportion of online education users in China from June 2018 to June 2021[12]

3. The Recovery of the Online Car-hailing Industry Accelerates

Under the impact of the COVID-19 pandemic, the demand for online car-hailing users dipped significantly in the first half of 2020. By June 2020, the scale of online car-hailing users in China dropped to 340 million, just equivalent to 84.1% of the same term in 2019. However, with the positive progress made in the prevention & control of COVID-19, the public demand for online car-hailing picked up in the second half of 2020, and the number of user rebounded rapidly. By June 2021, the scale of online car-hailing users in China had reached 397 million, with an increase of 56.4 million over June 2020, accounting for 39.2% of all Internet users. As unmanned driving technology matures, many policies have been formulated to support the development of the automatic driving industry. In February 2020, the National Development and Reform Commission and ten other ministries and commissions jointly released the *Smart Car Innovation and Development Strategy*, which drew the blueprint of realizing the mass production (L3) of conditional intelligent driving vehicles by 2025. On January 1, 2021, the *Classification of Auto Driving Automation* formally took effect, which specified the classification principles and technological requirements of the auto driving automation system. As one of the basic standards of the intelligent networked vehicle standard system in China, it provides support for the introduction of subsequent laws, regulations and mandatory standards related to auto driving in China. Beijing, Shanghai, Guangzhou, Wuhan and other cities slackened the regulation on the license of automatic driving and manned testing. As related policies

---

[12] Data Source: *The 48th Statistical Report on China's Internet Development* released by the China Internet Network Information Center.

required, the online car-hailing industry in China started to use new energy by the end of 2020 in an all-round way, improved car sharing rate, optimized urban transport capacity and reduced traffic pollutant emissions by innovating travel models and adjusting pricing strategies.

3.6 New Industries, Business Forms and Models Advance Towards High-Quality Development

Presently, finance and technology are deeply integrated, which deepens new financial infrastructure, independent innovation capacity and technological standards. E-commerce progresses steadily, and cross-border e-commerce, livestreaming e-commerce and social e-commerce develop rapidly. With sharing economy gradually segmented, the imbalance in the development of sharing economy in different fields becomes more prominent.

3.6.1 Fintech Continues to Deepen

Fintech entities accelerate market layout. The People's Bank of China and major banks and financial institutions speed up the strategic deployment of Fintech enterprises. According to the statistics of the *White Paper on China's Fintech Ecology 2020*, by August 2020, the People's Bank of China had successively established Shenzhen Fintech Co., Ltd., Yangtze River Delta Fintech Co., Ltd. and CFIT Co., Ltd. in Shenzhen, Suzhou and Beijing, which would play a leading role in the development of local industries. Five major state-owned banks and most joint-stock banks also own independent Fintech subsidiaries that enhance their core competitiveness in Fintech.

Key technological applications of Fintech quicken. Intelligent technologies are continually integrated, developed and applied, which boosts the intelligent transformation of the whole-process financial business. Currently, Fintech triggers far-reaching changes in such fields as front-end service and marketing, middle-end product and risk control, and back-end management and data of the financial industry. Blockchain technologies develop fast, and the use cases of financial blockchain continuously evolve in practice. Now, blockchain technologies are applied to supply chain finance, trade financing, payment and clearing, fund management and other segmented fields.

The pilot of e-CNY advances in an all-round way, and pilot cities and areas continuously expand. The pilot basically covers the Yangtze River Delta, the Pearl River Delta, Beijing-Tianjin-Hebei as well as central, western, northeastern and northwestern regions. By June 2021, more than 1.32 million pilot use cases were created, more than 20.87 million personal wallets were opened, and the accumulated transaction amount was *c.* 34.5 billion *yuan*.

3.6.2 E-commerce Develops Steadily

E-commerce transaction volume in China increased from 21.8 trillion *yuan* in 2015 to 37.2 trillion *yuan* in 2020.[13] In the past year, e-commerce made steady progress.

Cross-border e-commerce grows against headwinds. According to the statistics released by the General Administration of Customs, in the first quarter of 2021, the export volume of goods trade in China reached 4.61 trillion *yuan*, with an increase of 38.7%. Particularly, the import and export volumes of cross-border e-commerce realized 419.5 billion *yuan*, with a year-on-year increase of 46.5%. In terms of categories, traditional cross-border e-commerce export commodities in China are mainly distributed in the field of daily consumer goods at low prices. In recent years, the export of high-end manufacturing became a new trend, and "Made in China" advanced towards "Intelligent Manufacturing". China plays an active role in promoting international cooperation and development in the field of e-commerce. In November 2020, China formally signed the *Regional Comprehensive Economic Partnership* (RCEP) with ten ASEAN countries and the other four countries (i.e., Japan, South Korea, Australia and New Zealand), which clarified basic principles for each member party to formulate data localization and cross-border data flow policies in e-commerce projects. In June 2021, the Ministry of Commerce launched China-CEEC E-commerce Cooperation & Dialogue Mechanism. Six partner countries, i.e., China, Albania, Hungary, Serbia, Montenegro and Slovenia, strengthen exchanges and cooperation in the field of e-commerce and share the dividends of digital-economy development on the basis of the principle of openness, voluntariness and sharing.

Livestreaming-based e-commerce sees the stage of "survival of the fittest". In 2020, the scale of China's livestreaming-based e-commerce industry totaled 916 billion *yuan*, with a year-on-year increase of 111%.[14] Simultaneously, as the industrial threshold and professional requirement continue to rise and users become inured to livestreaming, some models are squeezed out. On the one hand, the form of "gaining sales by traffic" does not pan out. On the other hand, some multi-channel networks (MCN) without core competitiveness encounter business crises or even go bankrupt.

Social e-commerce integrates social function with e-commerce, and the trend of community group purchase and customer-to-manufacturer (C2M) becomes more prominent. According to the *2021 China Social E-commerce Industry Development Report* released by the Internet Society of China, in 2020, the overall scale of social e-commerce achieved 3.7 trillion *yuan*, the number of consumers approached 700 million, and the number of employees exceeded 70 million, with an integration degree of tradition and innovation of *c*. 70%.

---

[13] Data Source: *Digital China Development Report 2020* published by the Cyberspace Administration of China.

[14] Data Source: iiMedia Research.

3.6.3 *Sharing Economy is Gradually Segmented*

In 2020, under the impact of the COVID-19 pandemic, the overall market growth of sharing economy slowed down significantly, and the imbalance in the development of sharing economy in various fields intensified. The business revenue of the industries with high online & offline integration, such as shared accommodation and shared travel, reduced sharply. In contrast, industries such as shared medical care, shared education and community group purchase grew against headwinds in the pandemic, and sharing economy faced both development opportunities and challenges. As *China's Sharing Economy Development Report 2021* estimated, in 2020, the transaction scale of sharing economy market in China approximated to 3,377.3 billion *yuan*, with a year-on-year increase of *c.* 2.9%. The overall growth rate slowed down significantly over 2019. In terms of market structure, the transaction scales of sharing economy market in three fields, i.e., life service, production capacity and knowledge & skill, rank top three, reaching 1,617.5 billion *yuan*, 1,084.8 billion *yuan* and 401 billion *yuan* respectively. With respect to development speed, the three fields with the largest decline in transaction scale were sharing accommodation, sharing office and sharing travel, down by 29.8, 26 and 15.7% respectively year on year. Additionally, the transaction scale of the sharing economy in the life service field decreased by 6.5% year on year. Since the COVID-19 pandemic severely buffeted the catering industry, domestic service and offline leisure & entertainment, and the consumption on life services slipped significantly. In regard to job market, the COVID-19 pandemic enormously stimulated the employment demand in the fields of livestreaming & short video and knowledge sharing. In 2020, the Ministry of Human Resources and Social Security, together with relevant departments, released a list of 25 new occupations, e.g., intelligent manufacturing engineering and technology personnel, industrial Internet engineering and technology personnel and online delivery personnel.

3.7 The Pandemic Objectively Accelerates the Development of the Digital Economy

COVID-19 objectively expedites the development of new business forms and models in the digital economy. In the long term, the pandemic will quicken socio-economic digital transformation and the development of the digital economy. The demand for consumption and investment in the digital economy will continue to increase.

3.7.1 The Pandemic Speeds up the Development of New Business Forms and Models in the Short Term

The COVID-19 pandemic sets off a variety of new digital demands immediately. In the field of life, new digital forms and models such as teleworking, online education and online video sprung up. In the field of production, enterprises use big data and industrial Internet to realize the resumption of work and production, the accurate connection between supply and demand, efficient production, overall allocation and capital financing, etc. This breeds new models of digital production, accelerates industrial digital transformation, and deepens integrated development. In 2020, the proportions of the added value of the digital economy in China's service industry, industry and agriculture increased rapidly, up by 7.67, 7.69 and 8.54% respectively over 2019.

3.7.2 The Pandemic Speeds up the Digital Transformation in the Long Term

On the demand side, in the COVID-19 pandemic, digital transformation possesses great significance in strengthening the resilience of enterprises and arousing the willingness of enterprises and governments to adopt digital transformation. Enterprises that forge solid digital foundations use digital technologies to timely improve the industrial chain and hedge against the negative impact of the pandemic. According to the data of the China Association for Small & Medium Commercial Enterprises, the recovery rate of enterprises with high digital maturity reached as high as 60% within three months under the impact of the pandemic, whist that of enterprises with low digital maturity reached only 48%. On the supply side, the COVID-19 pandemic propels the construction and improvement of digital infrastructure and promotes the innovation and upgrading of digital products and services, thus enhancing and empowering the supply side. Simultaneously, the construction of new-type digital infrastructure like 5G, data centers and industrial Internet accelerates, which furnishes more high-performance digital tools and resources for the development of the digital economy and better supports or empowers the digital transformation of the economy and society.

3.7.3 The Pandemic Enlarges the Demands for Consumption and Investment in the Digital Economy

As the development and innovation of digital technology deepen, new digital products and services continue to emerge, and the digital-consumption field further expands, which meets the consumption needs of the public in a diverse way. The entities of digital investment also develop from a single party to multiple parties. In the digital economy, the gap between investments in hardware and service narrows down, and the investment in new-type infrastructure continuously accelerates in an all-round way. As a part of new-generation information technology, digital technology has become the key area of national scientific & technological strategies and corporate development strategies. In the future, the macro environment for the development of the digital economy will be optimized, with policies and systems more complete. This will provide environmental support for the healthy and sound development of the digital economy.

Chapter 4
The Development of E-government

4.1 Outline

The *Outline of the 14th Five-Year Plan for National Economic and Social Development* draws the grand blueprint for China to embark on a new journey of comprehensively building a modern socialist country. A special part concerns "improving the level of digital government construction", and proposes to strengthen the opening and sharing of public data, promote the joint contribution and shared benefits of information-based government and improve the efficiency of digital government services. In the past year, in terms of the construction and development of digital government and e-government, positive progress was made in the planning, layout and coordination of E-government, the supporting construction of information infrastructure, the sharing, opening and development of data resources, the application to government affairs, the construction of digital society and the COVID-19 prevention & control.

4.2 The Planning, Layout and Coordination of E-government Continue to Improve

Under the overall deployment of the CPC Central Committee and the State Council, various regions at all levels enthusiastically boost the construction and development of e-government. In particular, in the past year, a series of special policies were issued in relation to the construction and development of digital government, and relevant work was systematically arranged and coordinated with positive progress.

Chinese Academy of Cyberspace Studies, *China Internet Development Report 2021*,
https://doi.org/10.1007/978-981-19-9311-4_4

4.2.1 The Planning and Coordination of E-government is Carried Out in an All-Round Way

In 2021, the *Outline of the 14th Five-Year Plan for National Economic and Social Development* was released. The complete deployment of the "Digital China" Strategy was systematized, and the construction & development requirements of e-government were further clarified. The *Outline* proposes to improve the construction level of digital government. Digital technologies will be widely applied to government management & services, so as to promote the reconstruction of government-management procedures and the optimization of government-management models, and continuously improve the scientificalness of decision-making and the efficiency of service, by taking such measures as strengthening the opening and sharing of public data, facilitating the joint contribution and shared benefits of information-based government, and raising the efficiency of digital-government services. The construction of a digital society will be accelerated. In line with the construction requirements of smart and convenient public services, smart cities and digital villages, smart communities and digital families, digital technologies will be comprehensively integrated into social exchanges and daily life, so as to quicken the innovation on public services and social operation models and build a digital life that all citizens can enjoy. In brief, the *Outline* raises clear requirements for the construction and development of e-government in China.

At the local level, by June 2021, 31 provinces (autonomous regions and municipalities directly under the Central Government) and Xinjiang Production and Construction Corps had released the *Outline of the 14th Five-Year Plan for National Economic and Social Development and* the *Long-Range Objectives Through the Year 2035*. In combination with local conditions, various regions have made measures to advance the construction of the digital economy, digital society, digital government and digital ecology. Table 4.1 shows special plans in relation to recently launched digital government construction in various regions.

Under the background of implementing the "Digital China" Strategy, the construction, planning and layout of digital government at the central and local levels basically form a complete system. For the future development of digital government, there are four directions as deployed.

(1) Highlighting the content in the construction of digital government. On the basis of the existing overall planning and layout of e-government, governments at various levels fully take into account the status quo of digital government, focus on key tasks, build the digital core-system framework according to the basic functions of governments, and continue to improve the digital support capacity for governments to perform their duties. In content, the special part that concerns digital government in the *Outline of the 14th Five-Year Plan for National Economic and Social Development* at the provincial level includes public data sharing platform and data opening, "one-website handling" in government service, "one-website management" in government governance and "one-website coordination" in government operation,

Table 4.1 Special plans in relation to recently-launched digital-government construction in Various Regions

Regions	Documents	Document no	Date of issuing
Hubei	The master plan for digital government construction in Hubei Province 2020–2022	*E' Zheng Fa* [2020] No. 12	June 17, 2020
Shanxi	The plan for digital government construction in Shanxi Province 2020–2022	*Jin Zheng Ban Fa* [2020] No. 79	September 29, 2020
Anhui	The plan for digital government construction in Anhui Province 2020–2025	*Wan Zheng* [2020] No. 44	October 24, 2020
Henan	The master plan for digital government construction in Henan Province 2020–2022	*Yu Zheng* [2020] No. 35	December 11, 2020
Ningxia	the action plan for digital government construction in Ningxia Hui Autonomous Region 2021–2023	*Ning Zheng Fa* [2021] No. 11	March 1, 2021
Zhejiang	The "14th five-year plan" for digital government construction in Zhejiang Province	*Zhe Zheng Fa* [2021] No. 13	June 4, 2021
Guangdong	The "14th five-year plan" for the reform and construction of digital government in Guangdong Province	Yue Fu [2021] No. 44	June 30, 2021

as well as the construction of basic supporting capacity of digital government like government cloud, government network and security protection.

(2) More entities in the construction of digital government. Digitization serves as an important force that drives the overall transformation of production mode, lifestyle and governance model. In the *Outline of the 14th Five-Year Plan for National Economic and Social Development*, most provinces clearly take "digitization" as a development goal. In public services of education, medical care, elderly care, culture and sports, various governments place a high premium on joint contribution and shared benefits among multiple entities. In the intelligence-based field, market entities and public enterprises and institutions play an important part in promoting the integration and development of digital government, digital economy and digital society. The multiple-entity collaboration between governments, enterprises and institutions has become an important development model.

(3) More complete rules and systems. Overall digital transformation relies on an open, healthy and safe digital ecology. The construction of digital government plays a fundamental and leverage role in forging benign digital ecology. The scientific and reasonable open sharing of public data means the premise and foundation of realizing the optimization and allocation of data elements. Systems and regulations on the

protection of personal information and the protection & classification of data maintain the digital order. The construction of a critical information infrastructure protection system constitutes a key measure to safeguard cybersecurity and political security. To enhance regulatory capacity in the construction of digital government ensures the healthy and sustainable development of comprehensive digital transformation. Therefore, continuously improving the systems and rules on digital government boots the standardization and management of digital reform in an all-round way.

(4) More prominent penetration effect. Digital and intelligent applications penetrate into all areas of the national economy and social development. The construction of digital government calls for integrity and coordination. The statement "to make full use of digital technologies to improve governmental function and governance capacity" is reiterated in more extensive fields, such as foreign trade, energy, infrastructure, transportation, water conservancy, fiscal and taxation, business-environment optimization, finance, agricultural & rural construction, urbanization, cultural tourism, health, social security, education and grass-roots governance. With the support of a uniform platform of digital government, more efforts are taken to strengthen the coordination of governmental internal and external business and cross-department governmental-procedure optimization. Take the *Outline of the 14th Five-Year Plan for National Economic and Social Development* in Guangdong Province as an example. There are 147 times of "digital", 63 times of "smart" and 93 times of "intelligence" in the full text.

4.2.2 The Coordination Mechanism is Continuously Improved

1. The Central Government Attaches Great Importance to Optimizing Government Data Sharing and Coordination Mechanisms

Implementing the big data strategy is an important measure for the governance of China. The sharing and opening of government data plays a major role in developing big data and forms an important of building digital government. China pays high attention to and sets out to address the problem of the unsound coordination mechanism of government data sharing, so as to unclog the channel for the development of digital government at the national level.

In December 2020, the *Opinions on Establishing and Improving the Coordination Mechanism for Government Data Sharing and Accelerating the Orderly Sharing of Data* was reviewed and adopted at the 17th meeting of the Central Committee for Deepening Overall Reform. The meeting emphasized that to establish and improve the coordination mechanism for government data sharing and accelerate the orderly sharing of data, various governments should adhere to the principle of overall-coordination-based, application-based, security-control-based and laws-based development of data, focus on business collaboration, strengthen the innovation on technology, application and model, and comprehensively build security system, management system and technological-protection system for government data sharing, so as to break down information barriers among departments, promote more accurate and smooth data sharing, and raise the level of legalization, institutionalization and standardization.

In March 2021, in the *Report on the Work of the Government*, Premier Li Keqiang of the State Council urged efforts to "strengthen the construction of digital government, establish and improve the coordination mechanism of government data sharing, promote the expansion of the application of electronic licenses and national exchange and mutual recognition, and achieve more online, handheld and one-time government services".

2. Local Governments Lay Stress on Improving the Overall Planning and Coordination Capacity of Competent Authorities

The construction of digital government calls for coordination and linkage among multiple departments. It is a major measure to ensure the sustainable and healthy development of digital government to clarify the functional authorization and the establishment of rules and regulations of relevant departments and posts.

In April 2021, the General Office of Guangdong Provincial People's Government released the *Notice on Printing and Distributing the Plan for Pilot Work of the Chief Data Officer System in Guangdong Province*. The pilot work is carried out in six provincial departments (including Guangdong Provincial Department of Public Security and Guangdong Provincial Department of Human Resources and Social Security) as well as ten cities at the prefecture level (including Guangzhou, Shenzhen and Zhuhai), to establish the chief data officer system and deepen the reform of the market-oriented allocation of data elements. The responsibilities of the chief data officer mainly include four aspects.

(1) Promoting the construction of digital government, implementing the deployment of superior departments, and organizing the preparation of digital government development plans, standards and specifications and implementation plans.
(2) Coordinating data management, integration and innovation, organizing the preparation of medium & long-term development plans and relevant system specifications for data governance, coordinating internal and external data needs, and promoting data sharing, opening, development and utilization.
(3) Implementing regular guidance and supervision, coordinating and solving major problems in the construction of informatization projects, and supervising the implementation of data governance operation and informatization construction.
(4) Strengthening the construction of a talent team, promoting the construction of data operation organization, data governance and operation team at the same level, and organizing data skill & security training.

Over the past year, various regions continuously implemented relevant requirements of *National Measures for the Construction and Management of Government Information-Based Projects* and deepened overall management of government-informatization-construction projects. Jiangsu, Jiangxi, Hainan, Guangxi, Jilin and other provinces successively introduced relevant policies to accelerate the intensive construction of government-information infrastructure, the interconnection of government information systems, the sharing of government data resources and business collaboration, and improve the application performance of government information systems. Table 4.2 shows relevant documents on the management of government information-based-project management in some regions.

4.2.3 Relevant Policies Are Continually Promulgated

1. Promoting the Realization of Services for Cross-Governmental Affairs

The CPC Central Committee and the State Council continue to advance "Internet + Government Services". The National Integrated Government Service Platform takes the initial shape and plays a significant role. On the basis of popularizing "one-website handling" in government service, at the end of September 2020, the *Guiding Opinions of the General Office of the State Council on Accelerating the Promotion of "Services for Cross-Governmental Affairs"* was formulated and issued. It took painstaking efforts to address the conundrums in government services that troubled citizens, when they had to handle affairs in different regions many times, requiring that 140 high-frequency inter-provincial government services would be realized by the end of 2021. By deepening the whole-process website handling, expanding the agency services in different regions and optimizing the collaboration in multiple regions, governments can meet the needs of various market entities and the masses in handling affairs in different regions to the maximum extent.

2. Narrowing the Digital Divide and Helping the Elderly to Use Intelligent Technologies

In order to further promote the solution to the digital divide among the elderly who have difficulty using intelligent technologies and enable the elderly to better share the achievements of IT development, in November 2020, the General Office of the State Council released the *Notice on the Implementation Plan for Effectively Solving the Difficulties of the Elderly in Using Intelligent Technology*. It centers on seven high-frequency areas and service cases in the daily life of the elderly (e.g., travel, medical treatment, consumption, entertainment and socializing), combines the traditional service model with intelligent service innovation, and puts forward 20 measures and requirements.

Besides, the *Notice* regards bridging the digital divide of the elderly as a direction for the construction of digital government in the future, and brings forward scientific and reasonable working measures.

Table 4.2 Relevant documents on the management of government information-based-project management in some regions

Regions	documents	Document no	Date of issuing
Jilin	Notice of Jilin provincial administration of government affairs service and digital construction, Jilin provincial development and reform Commission, and Jilin provincial department of finance on printing and distributing the administrative measures for the construction of government affairs information-based projects in Jilin Province	*Ji Zheng Shu Lian* [2020] No. 9	September 10, 2020
Jiangxi	Notice of the general office of Jiangxi provincial people's government on printing and distributing administrative measures for the construction of government affairs information-based projects in Jiangxi province	*Gan Fu Ting Zi* [2020] No. 68	September 25, 2020
Hainan	Notice of the general office of Hainan provincial people's government on printing and distributing the administrative measures for the construction of government affairs information-based projects in Hainan Province	*Qiong Fu Ban* [2020] No. 38	December 13, 2020
Guangxi	Notice of the general office of Guangxi Zhuang autonomous region people's government on printing and distributing the administrative measures for the construction of Guangxi government information-based projects in Guangxi (for trial implementation)	Gui Zheng Ban Fa [2021] No. 21	March 24, 2021
Jiangsu	Notice on printing and distributing the administrative measures for the construction of provincial government information-based projects in Jiangsu Province	Su Zheng Ban Fa [2021] No. 24	April 27, 2021

(1) Further deepening the automated application of advanced technologies. For example, strengthening the application of AI, face recognition and other technologies, developing from "code scanning" recognition to "face scanning" recognition, and using intelligent functions to assist the elderly in their life.
(2) Guiding the elderly-friendly transformation of service terminals and service interfaces. For example, launching mobile terminals with large screens, large fonts, large volumes, large battery-capacity and simple OS, with various application software better adapting to such terminals. "One-Click Car-Hailing" is provided by online car-hailing software to simplify the operation to make it easy to use for the elderly.
(3) Maintaining and improving telephone service and other functions, strengthening the integration of various hotlines, and raising the response rate and service quality.
(4) Attaching more attention to the integration of online and offline services. For high-frequency services such as medical care, social security, civil affairs, finance, telecommunications, postal service, public complaints and proposals, entry-exit and living expenses, offline processing channels are reserved and extended to the grass-roots level to provide convenient on-site escort services for the elderly.

4.3 The Construction of Supporting Government Information Infrastructure Continues to Advance

Advanced infrastructure and system platforms support the implementation of the development strategy of "Digital China", and possess great significance for realizing the digital upgrading and transformation of various industries, improving the convenience of various applications in the digital society, and enhancing the governance and service capacity of digital government. Over the past year, local governments at various levels sped up the planning and layout of new-generation information infrastructure and continued to build e-government-related systems and platforms in various fields and industries, with remarkable achievements.

4.3.1 The Planning and Layout of Infrastructure Are Strengthened

In December 2020, the National Development and Reform Commission and other departments jointly issued the *Guiding Opinions on Accelerating the Construction of National Integrated Big Data Center Collaborative Innovation System*, which clearly stated that by 2025, the public cloud service system would be basically formed, and the cost of obtaining computing power in the whole society would be significantly reduced. Besides, the *Opinions* proposed to "build an integrated computing service system" and carry out systematic planning and deployment, which would further promote the intensive construction and application of infrastructure in the E-government field and improve the efficiency of construction and management.

In November 2020, the General Office of Guangdong Provincial People's Government released the *Notice on Printing and Distributing the Three-Year Implementation Plan for Promoting the Construction of New-Type Infrastructure in Guangdong Province 2020–2022*, which proposed to actively deploy high-quality 5G networks, explore & build special wireless government network based on 1.4 GHz frequency band, construct government big data centers, provide domestic government cloud services, support the construction of "city brain" in eligible cities like Guangzhou and Shenzhen, advance the integration of urban-management public facilities with 5G, IoT and sensing technologies, and deploy smart-sensing nodes for urban data collection.

In December 2020, the General Office of Shanghai Municipal People's Government issued the *Notice on Printing and Distributing the Action Plan for the Construction of New-Type Community Infrastructure* to promote the comprehensive integration of 5G, AI, IoT, big data and other new technologies into community life, optimize the system of smart community, make community governance more intelligent and improve the quality of community life. As deployed, major tasks are to accelerate the deep coverage of 5G, gigabit optical-fiber network and new-type metropolitan IoT in communities, implement the "community cloud" application platform in residential communities under a "one-website management" framework, improve intelligent-terminal-distribution facilities in communities, strengthen the intelligent security of communities, promote the construction of smart parking, smart energy, smart elderly care, smart health and other facilities in communities, encourage cooperative banks to increase loans at preferential interest rates for community new-type infrastructure projects, and provide discount-interest support for eligible community new-type infrastructure projects.

4.3.2 The Construction of the System Platform is Continuously Promoted

According to the analysis of the bidding documents released at www.ccgp.gov.cn from July 2020 to June 2021, 31 provinces (autonomous regions and municipalities directly under the Central Government) and Xinjiang Production and Construction Corps (hereinafter referred to as "The Corps") have carried out projects on digital-government platform & system construction (including services). To be specific, there are 651 platforms and system-construction projects related to provincial government-data governance, as shown in Fig. 4.1. Shandong (67), Beijing (52) and Guangdong (49) rank top three in the number of the construction project.[1]

[1] Data Source: http://www.ccgp.gov.cn.

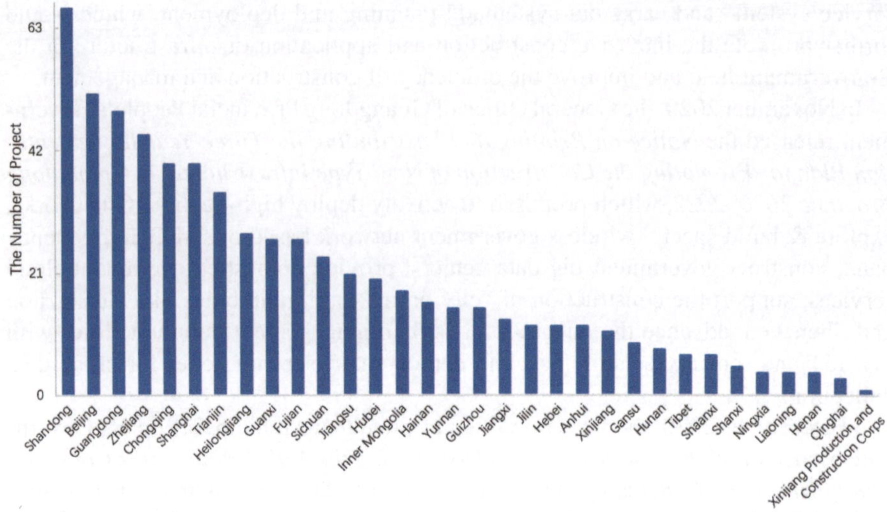

Fig. 4.1 The number of platform and system-construction projects related to provincial government data governance from July 2020 to June 2021

In terms of investment, as the bidding information at www.ccgp.gov.cn suggests, from July 2020 to June 2021, the cumulative investment to construction & service projects related to provincial government data was 3,840,423,500 *yuan*. Top three provinces (municipalities directly under the Central Government) were Zhejiang (582,269,400 *yuan*), Tianjin (521,901,200 *yuan*) and Beijing (388,232,500 *yuan*) in investment, as shown in Fig. 4.2.

4.3.3 The Construction of Provincial Government Data in Various Fields is Constantly Enhanced

From July 2020 to June 2021, the construction & service projects related to provincial government data mainly involved 27 fields, such as comprehensive application, public security, environment, finance & taxation, health & family planning and industry & information technology. The number of projects in the comprehensive application was the largest, reaching 148 and accounting for 22.7% of the total. The data comprehensive application projects were mainly "Internet + government service" and data sharing & opening built by data-comprehensive-management departments like local big data bureaus and general offices of people's governments. In the fields of public security and environment, the number of projects in public security reached 54 and accounted for 8.3% of the total, and the number of projects in the environment reached 51 and accounted for 7.8% of the total. Besides, the number of projects in health & family planning, finance & taxation and industry &

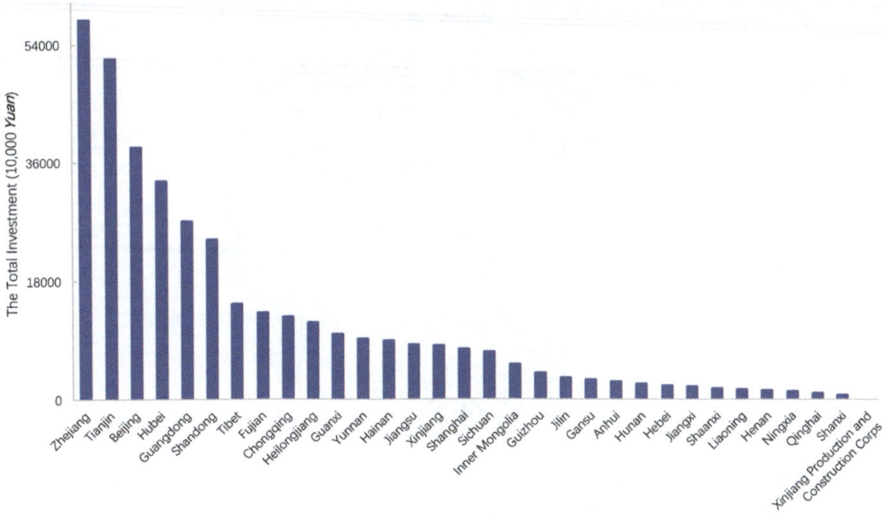

Fig. 4.2 The total investment to relevant construction & service projects related to provincial government data governance from July 2020 to June 2021

information technology all exceeded 30, reaching 45, 42 and 34 respectively, with a high proportion. Figure 4.3 shows the number of construction & service projects related to provincial government data in various fields from July 2020 to June 2021.

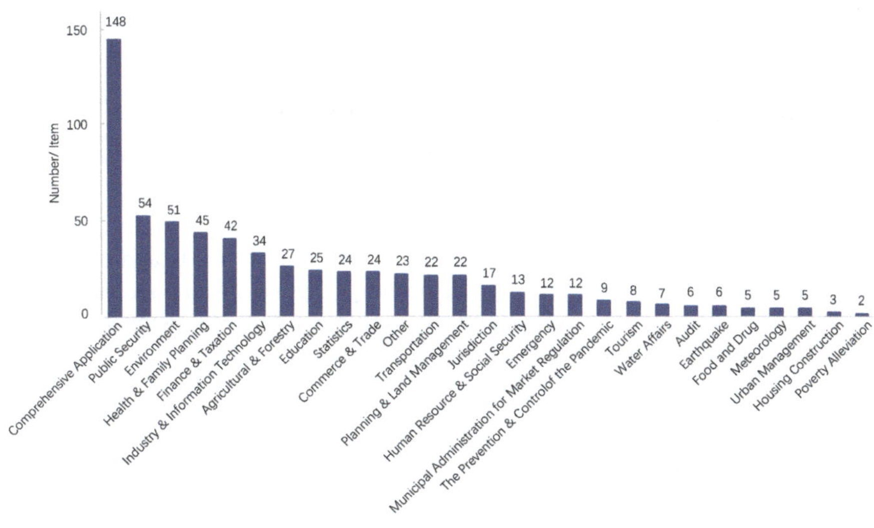

Fig. 4.3 The number of construction & service projects related to provincial government data in various fields from July 2020 to June 2021

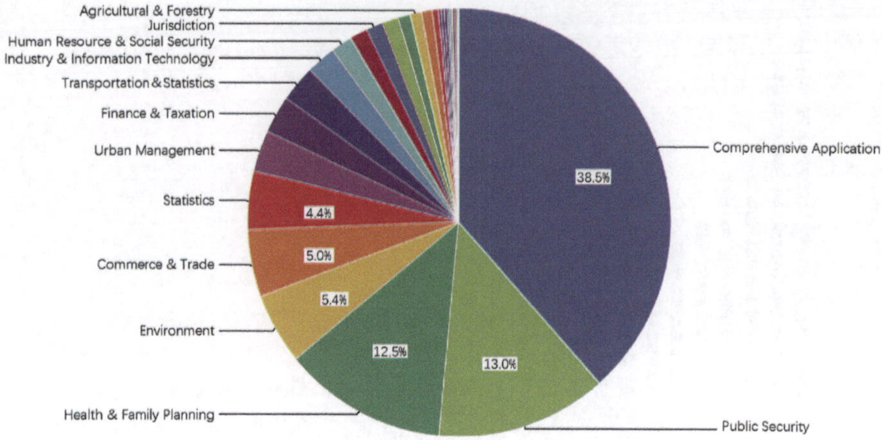

Fig. 4.4 The capital investment to construction & service projects related to provincial government data in various fields from July 2020 to June 2021

In capital investment, the investment in the construction of comprehensive applications was the largest, amounting to 1.479 billion *yuan* and accounting for 38.5% of the total. Public security, health & family planning and environment amounted to 498,413,400 *yuan*, 479,224,700 *yuan* and 206,860,800 *yuan* and accounted for 13.0, 12.5 and 5.4% of the total respectively. Figure 4.4 shows the capital investment in construction & service projects related to provincial government data in various fields from July 2020 to June 2021.

4.4 The Sharing, Opening and Development of Data Resources Continue to Deepen

The sharing and opening of government data serve as a key link to deepen e-government and build a digital government. Various departments and regions in China continue to extend the sharing and opening of government data and deepen the sharing, opening and development of government data via channels such as sharing exchange platforms and data opening platforms.

4.4.1 Data Sharing

With the advancement of data sharing, data query & verification are extensively applied to data sharing, with the advantages of data freshness and service timeliness. As of May 2021, the National Data Sharing and Exchange Platform had more than

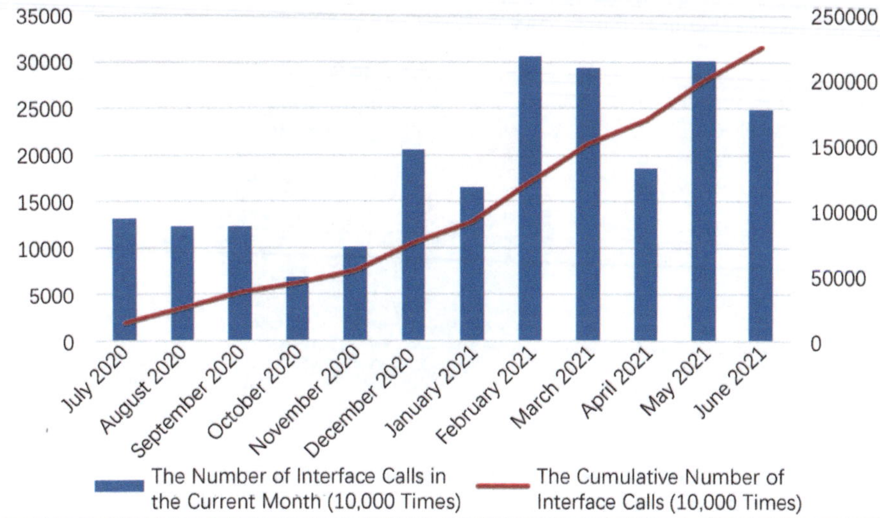

Fig. 4.5 The interface calls of data query & verification of the national data sharing and exchange platform from July 2020 to June 2021

650,000 online directories and more than 1,200 shared interfaces. Since its launch, the platform has provided data query & verification services more than 3.7 billion times. From July 2020 to June 2021, the maximum number of data query & verification services via the platform reached 306 million monthly, with a total of more than 2.2 billion times, as shown in Fig. 4.5. *Vis-à-vis* the data before July 2020, the number of monthly average data query & verification interface calls increased from 80 million times to more than 180 million times, with remarkable data sharing effect.[2]

At the national level, the demand for government data sharing remains a high concentration ratio. From July 2020 to June 2021, five departments, i.e., the Ministry of Public Security, the Ministry of Education, the State Administration for Market Regulation, the General Office of the State Council and the Ministry of Civil Affairs, provided 2.1 billion times of data query & verification services in total, accounting for more than 94% of all data query & verification services on the National Data Sharing and Exchange Platform. In particular, the demands for data sharing rank the top in the basic information of natural persons of the Ministry of Public Security, information on higher education degrees of the Ministry of Education, basic information of enterprises of the State Administration for Market Regulation, and information of civil marriage registration of the Ministry of Civil Affairs, as shown in Fig. 4.6. Since May 2020, the pandemic-prevention health code information of the General Office of the State Council has been intensively used, which provided strong support for COVID-19 prevention & control in China.

[2] Data Source: National Data Sharing and Exchange Platform.

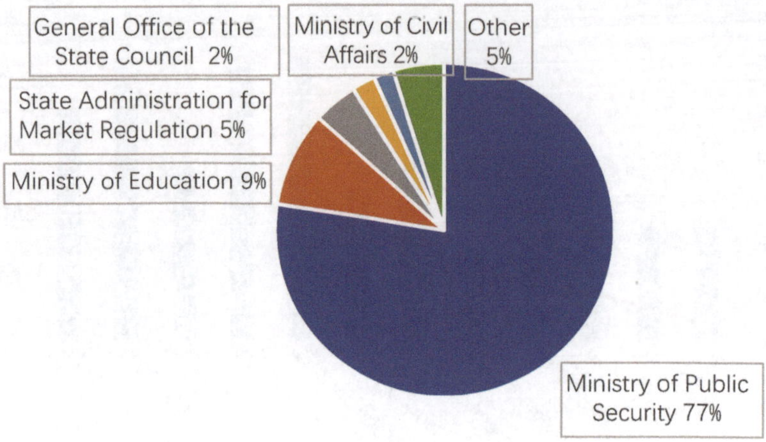

Fig. 4.6 The proportion of data query & verification services provided by the national data sharing and exchange platform from July 2020 to June 2021

In general, various regions need to use a substantial number of data from ministries and commissions. From July 2020 to June 2021, Zhejiang, Guangdong, Jiangsu, Yunnan, Liaoning, Hainan, Shandong, Shanghai, Shaanxi and Heilongjiang ranked top ten in the number of data query & verification via the National Data Sharing and Exchange Platform, as shown in Fig. 4.7. The demand for data acquisition was mainly distributed in the pandemic-prevention health code, higher education degree, professional qualification certificate, corporate basic information, and so on.[3]

In the data sharing and interaction among ministries and commissions, data sharing chains are fully built and data sharing services are widely actualized in some businesses. Among them, the typical one is the inquiry & verification of basic information of natural persons conducted by the State Taxation Administration in the special additional deduction business of personal income tax. The data sharing volume in this regard remains on the top of inter-departmental data sharing volume on the National Data Sharing and Exchange Platform, and the monthly average number of calls exceeds 66 million. Additionally, the Publicity Department of the CPC Central Committee carries out basic information query & verification of natural persons when conducting online game anti-addiction real-name authentication. The Ministry of Human Resources and Social Security carries out information query & verification of college education information when conducting the skill-qualification examination. The National Development and Reform Commission carries out basic information query & verification of enterprises when conducting administrative examination and approval. In these cases, the number of calls to the data sharing interface is also large.

[3] Data Source: National Data Sharing and Exchange Platform.

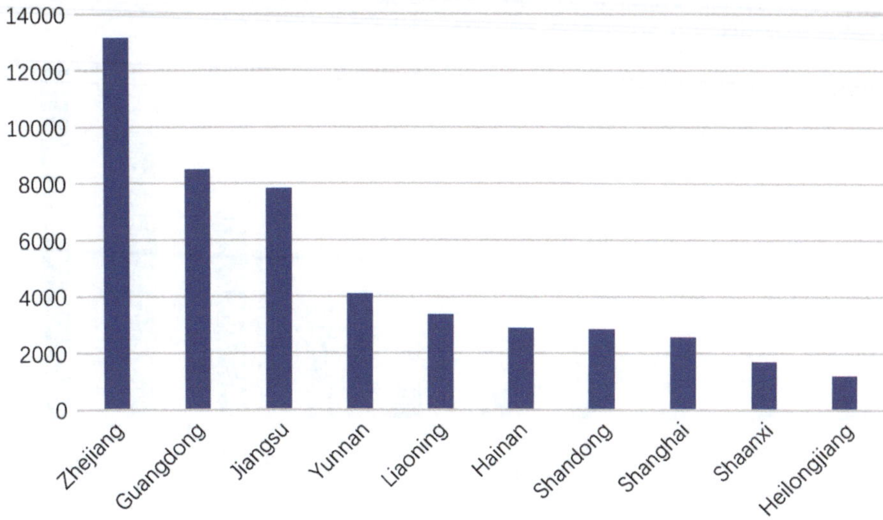

Fig. 4.7 Top ten provinces (municipalities directly under the central government) in the number of calls to data query & verification interface (10,000 times) on the national data sharing and exchange platform from July 2020 to June 2021

4.4.2 Data Opening

By April 2021, 174 provincial governments and municipal governments in China had launched data opening platforms, as shown in Fig. 4.8. The number of open government data platforms at and above the prefecture level increased significantly, from 20 in 2017 to 174 in the first half of 2021. The open government data platform increasingly becomes a typical model in the construction of local digital government and the governance of public data.[4] Figure 4.8 shows the changes in the number of local open government data platforms in China from 2012 to the first half of 2021.

Specifically, there were 18 provincial platforms (including provinces and autonomous regions, excluding municipalities directly under the Central Government and Hong Kong, Macao and Taiwan), 11 sub-provincial platforms and 141 prefecture-level municipal platforms, as shown in Fig. 4.9. Compared with the second half of 2020, 32 local platforms were added, and the total number of platforms increased by more than 20. 56.25% of provincial administrative regions (excluding Hong Kong, Macao and Taiwan), 73.33% of sub-provincial administrative regions and 44.34% of prefectural & municipal administrative regions launched open government data platforms.

[4] Data Source: *China Local Government Open Data Report* released by DMG Lab, Fudan University, by the first half of 2021.

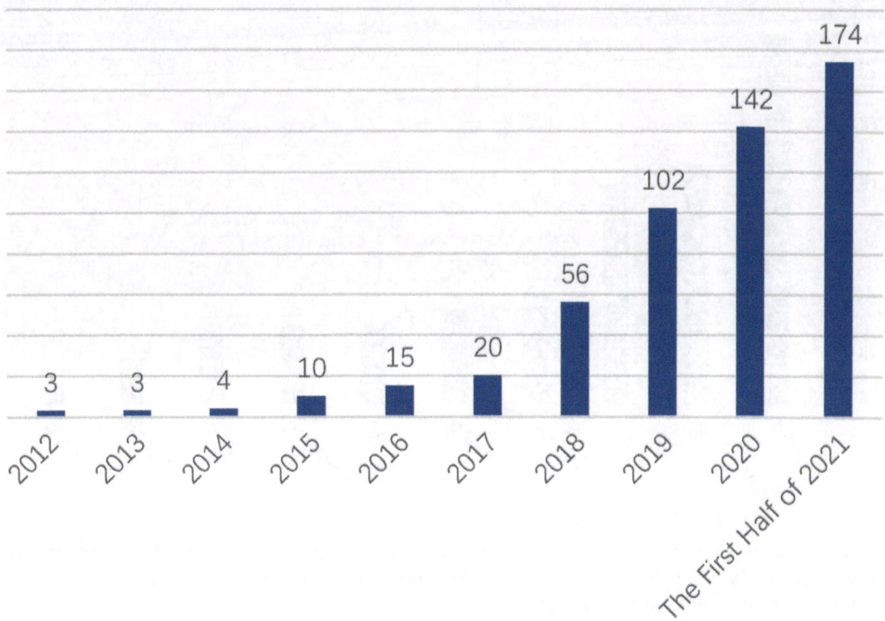

Fig. 4.8 The changes in the number of local open government data platforms in China from 2012 to the first half of 2021

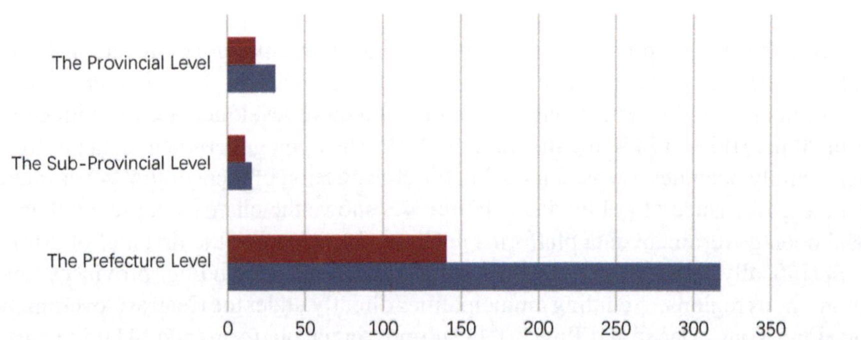

	The Prefecture Level	The Sub-Provincial Level	The Provincial Level
The Number of Platforms	141	11	18
The Total Number of Local Open Government Data Platforms	318	15	31

Fig. 4.9 The number of local open government data platforms at various levels in China by the first half of 2021

4.4.3 *The Development and Utilization of Data*

The development and utilization of government data mean an important link for governments to improve work efficiency, service level and governance capacity in the digital era. In recent years, local departments attached great attention to the value of data elements, continued to deepen the development and utilization of data, and energetically promoted the release of the value of government data. Take Beijing and Fuzhou as examples.

Beijing Financial Public Data Zone signals an innovative practice of data opening, which aims to strengthen the application of public data to the financial and social fields, foster the development of inclusive finance, improve the business environment and accelerate the construction of the smart city. In September 2020, the Beijing Municipal Bureau of Economy and Information Technology and Beijing Financial Holdings Group signed the *Agreement on Authorized Operation and Management of Beijing Financial Public Data Special Area* to explore ways to promote the application of financial public data by means of authorization and opening. By December 2020, Beijing Financial Public Data Special Area had collected 3,369 pieces of high-value data from big data platforms in Beijing, in 258 categories (including industrial & commercial registration, tax payment, social security, jurisdiction, housing provident fund, real estate, patent and governmental procurement) from 28 organizations that covered more than two million market entities, involving around 1.7 billion pieces of data. Public data services had achieved full coverage of 22 banks settled in the first-loan service center. A total of 12,800 corporate financing businesses were supported and accepted, and the financing and credit-granting needs of 7,607 enterprises were considered and met, with financing services for 4,101 enterprises, totaling 23 billion *yuan*.[5]

In recent years, Fuzhou actively explored and built big data integration & innovation application project featuring "orderly gathering, unified management, deep sharing and efficient utilization", and established the "Fuzhou model" of "data processing". Fuzhou endeavors to build a number of data platforms and basic databases, gather rich "data resources" and introduces a series of management measures to protect the opening and sharing of data. Simultaneously, Fuzhou implements a set of data assessment systems to realize "promoting construction via evaluation", builds big data industrial park and industrial ecology, and explores the formation of a five-part data-asset-operation industrial chain of "data convergence + data governance + data opening + data application + data transaction". In this way, data assets become a new engine for the economic development of Fuzhou in the future.[6]

---

[5] *Beijing Daily*: "The Small and Micro Financial Service Platform Achieves Full Coverage in the City". December 5, 2020, https://news.sina.com.cn/c/2020-12-05/doc-iiznezxs5300843.shtml.

[6] www.fznews.com.cn: "Fuzhou Promotes the Development and Utilization of Data Resources to Empower High-Quality Development". December 14, 2020, https://baijiahao.baidu.com/s?id=1,686,000,776,136,855,194&wfr=spider&for=pc.

4.5 The Level of Government Data Application Continuously Improves

Presently, it is a hot topic among all sectors of society to quickly improve national and local integrated government service capacities, deeply promote "one-website management", constantly explore "innovative application", effectively empower digital transformation & development and regional digital governance, and better meet new expectations of people for a better life.

4.5.1 One-Website Handling

Grounded in the National Integrated Online Government Service Platform, Zhejiang takes into account user experience, grass-roots needs and reform tasks, comprehensively advances the construction of "One-Website Handling" 2.0 for government services, and initially establishes "Zhejiang Model" of "One-Website Handling" that features "handling affairs on one-website", "handling affairs at the nearest service halls" and "handling affairs in a more convenient way".[7] At the beginning of 2021, Zhejiang basically completed the transformation and upgrading of "One-Website Handling" 2.0 for applying for government services. Firstly, promoting institutional innovation. Zhejiang builds an efficient coordination mechanism that covers 61 provincial departments, 11 cities with districts and 101 office halls. Secondly, promoting technological innovation. Zhejiang vigorously builds a provincial integrated government service center, initiates the service model, and reinforces the support for the "One-Website Handling" platform. Thirdly, promoting management innovation. Zhejiang continues to strengthen digital empowerment, accurately obtains the handling of provincial government affairs and systematic operations in real time, and provides data support for accelerating service efficiency, government-function transformation and institutional reform. Fourthly, promoting service innovation. Zhejiang comprehensively standardizes the experience at four ends of government services (PC end, mobile end, window end and self-service machine end), constantly simplifies forms and reduces materials, actively explores the intellectualization of approval services, and continuously improves the convenience of handling affairs.

On the basis of the National Integrated Online Government Service Platform, Anhui further expands service contents, diversifies service channels and strengthens

[7] *Xiaoxiang Morning News*: "Zhejiang Ranks the 1st in China in Provincial Integrated Government Service Capacity". May 28, 2021, https://baijiahao.baidu.com/s?id=1700959410479979504&wfr=spider&for=pc.

platform functions to provide enterprises and citizens with full-coverage, non-discrimination and high-quality government services and social services.[8] This materializes important measures and practices for Anhui to upgrade government services and innovate the service model of digital government. By June 2021, mobile-end "Wan Shi Tong" App load reached 84 million times, with more than 8,100 services accessed.[9] In order to optimize online service experience, Anhui also launches the flow chart of government services to realize "one chart guiding affairs handling". Anhui re-organizes the contents, functions and other resources of provincial government services and makes innovation in this regard. Anhui starts to use the service map with both spatial location and service items, via which citizens can inquire about government services and handle affairs. By January 2021, the Anhui government-service map had collected 23,000 service places at all levels, *c.* 6,000 self-service terminals, and 1.91 million service items in the province. The average daily visit reached around 393,000, with a satisfaction rate of 99.6%.[10]

4.5.2 One-Website Management

In his inspection of Shanghai in 2019, General Secretary Xi Jinping stressed that we should grasp the key to the work of "One-Website Handling of Government Services" and "One-Website Management of Urban Operation". Local governments actively implement the important instructions and requirements of General Secretary Xi Jinping, and conduct a lot of practical exploration in "One-Website Management".

In 2020, Shanghai proposed the vision of "getting information with one screen and managing the whole city with one website". Centering on five dimensions, i.e., urban dynamics, urban environment, urban transportation, urban security & supply and urban infrastructure, Shanghai designs 86 first-level indicators, integrates 33 specific topics from 22 organizations, and initially builds an urban operation system (version 1.0) for urban operation management and emergency response.[11] On this basis, Shanghai focuses on strengthening the "six-one support" of "one map for governance elements, one network for interconnection, one pool for data convergence, one cloud for the urban brain, one platform for system development and one

---

[8] News 360: "Anhui Provincial: 'Wan Shi Tong' App on Five Ends is Officially Launched". September 12, 2020, http://www.xinwen360.net/article/20200912215379064.html.

[9] Anhui Provincial Bureau of Data Resources Management: "Letter to Reply to the Opinions of No. 929 Delegates of the 4th Session of the 13th Anhui Provincial People's Congress". June 6, 2021, http://sjzyj.ah.gov.cn/public/7061/40436726.html.

[10] Sohu News: "Press Conference on Deepening Simplification of Administration and Decentralization of Power and Continuously Optimizing Business Environment". January 22, 2021, https://www.sohu.com/a/446190884_120,207,614.

[11] *China Youth Daily*: "Shanghai Urban Transport System Realizes 'Getting Information with One Screen and Managing the Whole City with One Website'". April 14, 2020, http://app.why.com.cn/epaper/webphone/qnb/html/2020-04/14/content_102895.html.

portal for mobile applications", and optimizes & & upgrades "one-website management" version 2.0. In order to comprehensively promote "one-website management" work, Shanghai sets up the municipal-level urban operation & management center ("urban operation center"). As the entity of "one-website management", "urban operation center" effectively guides the coordinated development of the city via the innovation on systems and mechanisms. In practice, the "urban operation center" does not replace or assume the daily operation and management functions of relevant departments. Rather, it promotes business collaboration and data integration through the backbone model of the city brain.

Tianjin deeply investigates the needs of urban governance, and continuously expands the use cases of "Jin Xin Ban" (one-website handling) and "Jin Zhi Tong" (one-website management), which can be linked. The integration of the websites of government services and urban governance empowers the digital transformation of Tianjin.[12] For example, if citizen's "Health Code" at the "Jin Xin Ban" platform is identified as "red code" or "yellow code", their personnel information will be automatically pushed to the "Jin Zhi Tong" platform, which forms a closed loop of "big data + grid" governance. "Jin Xin Ban" and "Jin Zhi Tong" have the function of "taking photos at convenience". Therefore, when citizens find problems, they can upload problems in all places of the city at any time, which will be sent to the "Jin Zhi Tong" platform in 30 s. More than 238 affairs in nine categories and 38 subcategories can be handled on "Jin Zhi Tong". With more than 37,000 full-time (or part-time) grid members, 6.13 million social governance events are reported and 5.9 million of them are settled.[13]

4.5.3 Innovation and Application

Against the backdrop of accelerating the construction of digital government, it is an important topic and significant practice of national governance and social governance to realize new decision-making and coordination and make new progress by technological innovation, model innovation and scene innovation.

The Supreme People's Procuratorate actively explores new applications of "intelligent procuratorial work". In recent years, the Supreme People's Procuratorate vigorously reinforced the construction of an intelligent voice cloud platform for procuratorial organs. Presently, the Supreme People's Procuratorate and procuratorial organs in Tianjin, Zhejiang, Anhui, Xinjiang, Inner Mongolia and other regions have completed the construction of intelligent voice cloud platforms, which provide voice recognition, transcription, bilingual translation and other functions for procuratorial cadres and policemen, and improve work efficiency. Procuratorial organs

[12] Tencent: "The Big Data Case in Tianjin Wins '2021 Digital Government Excellence Achievement Award'". May 27, 2021, https://new.qq.com/rain/a/20210527A0DOGQ00.

[13] www.tjcac.gov.cn: "Tianjin Governs the City via Big Data and Builds Two Platforms of 'Jin Xin Ban' and 'Jin Zhi Tong". April 21, 2021, https://www.tjftz.gov.cn/contents/5992/333688.html.

in Shanghai, Jiangsu, Guizhou and other regions organize the R & D of big data judicial case handling assistance system, and use big data technology to embed the unified evidence standards into the case handling procedures. With data analysis & comparison, the system provides an accurate basis for case-quality evaluation.

The State Administration for Market Regulation established national unified electronic business license system. Electronic business license has been applied to the whole-process electronic registration, online annual report, official seal engraving, taxation, social security, provident fund, real estate registration, public resource transaction, tobacco license, bank-account opening and other matters. Various market entities can use the mobile terminal installed with electronic business license apps to show their business licenses and verify their identities online at any time. Now, national market entities have downloaded and used electronic business licenses to handle identity authentication, online login, electronic signature and other services more than 70 million times, which tremendously improves the service level of the corporate startup.[14]

In recent years, the Ministry of Culture and Tourism launched the construction of a comprehensive monitoring and emergency command platform. The platform comprehensively integrates massive dynamic data of cultural undertakings, cultural industries and tourism, establishes data connectivity system that covers national, provincial & municipal and grass-roots levels, and analyzes tourist behaviors based on AI, in a bid to create and form an intelligent analysis brain with data video as the core. This actively boosts the deep integration of Internet-technology innovation with the reform & development of culture and tourism.[15]

Shanghai implements the working deployment on establishing a "good or bad evaluation" system for government services, and effectuates all-channel evaluation, all-item evaluation and three-corresponding evaluation related to the policy of "promoting construction via evaluation", in order to realize the closed-loop management of evaluation, feedback, rectification and supervision.[16] By the beginning of 2021, more than 21 million pieces of evaluation data that involved 16 districts and 56 line departments had been effectively processed, and 10 categories of major indicators and 83 categories of detailed indicators had been used online to conduct an in-depth analysis on the evaluation data. More than 6,000 pieces of poor evaluation processing tasks have been sent to Shanghai "12,345" platform.

Fujian stages "Min Zheng Tong", a unified platform for provincial government service mobile office, and energetically build "Ma Shang Jiu Ban", a handheld digital

---

[14] www.zggxkjw.com: "China Has Built National Unified Electronic Business License System". May 3, 2021, http://paper.chinahightech.com/pc/content/202105/03/content_42,531.html.

[15] www.wenlvzhisheng.com: "Vigorously Upgrading Information Technology to Solve Knotty Problems in the Development of Cultural Tourism". June 17, 2020, http://www.cac.gov.cn/2020-06/17/c_1,593,958,026,032,792.htm.

[16] Shanghai Municipal People's Government: "Working Planning on Deepening the Reform of 'One-Website Handling' and Building a Comprehensive Service System". March 16, 2021, https://www.shanghai.gov.cn/nw12344/20210316/0c634a9e96584303b44d7d79f1c30e9e.html.

government platform.[17] With an information-based innovation environment, the platform innovates the government office model with Internet thinking, promotes the "screen" office, realizes full coverage of the CPC Committee, People's Congress, Government and CPPCC at the provincial level, and connects administrative organs at the provincial, municipal, county, town and village levels. After mobile transformation, the provincial collaborative office system, online office hall, "Internet + supervision system" and other platforms are realized on "Min Zheng Tong", as a new engine for government's digital transformation.

4.6 The Supporting Capacity of Digital Society is Continuously Enhanced

In recent years, as IT continued to improve government-governance efficiency and promote innovation on the government-service model, e-government deepened and raised the accuracy of social governance and the efficiency of public services in a sustainable way, and digital society quickly took shape. In July 2020, the Cyberspace Administration of China released the *Digital China Development Report 2020*, which revealed that digital society accelerated the formation of a pattern of extensive consultation, joint contribution and shared benefits and that digital services for the benefit of people continually met the needs of people for a better life.

4.6.1 Digital Governance Capacity is Constantly Raised

1. Positive Progress is Made in the Construction of Information-Based Emergency Management

The emergency management department builds an "emergency command map" by getting access to the data of the Ministry of Water Resources, Meteorological Administration and other institutions, which realizes the visual display of basic information, comprehensive information, dynamic resources and force scheduling of disasters and accidents. The emergency information reporting system achieves the full coverage of emergency departments at the four levels of ministry, province, city and county. The "Wang Luo Tian Xia" system captures information related to Internet disasters and accidents in real time. The "e-button" and intelligent outbound call system for disasters and accidents unblock the reporting channels for the public and disaster information assistants. The monitoring and early warning system for the safety production risk of hazardous chemicals is widely applied, which realizes

[17] www.fj.sina.com.cn: "The First Unified Mobile Office Platform in Fujian Province Based on the Information and Innovation Environment, 'Min Zheng Tong' Office App Was Launched". April 26, 2021, http://fj.sina.cn/news/2021-04-27/detail-ikmyaawc2038184.d.html.

networked monitoring of major hazard sources, and effectively improves the systematic risk-prevention capacity of hazardous chemicals. Simultaneously, with the use of "blockchain + QR code" technology, the emergency management department actualizes the "one code for one thing" detailed management, whole-process supervision and online traceability of national emergency materials. With the aid of supercomputing capacity, population early warning models and satellite remote-sensing image-recognition technology are used to effectively raise the emergency-management disaster-verification capability. The estimation and statistics of the disaster-caused transferred population, the centralized resettlement population as well as the number of flooded houses can be completed in a few hours, which requires a week in a traditional way.

2. The Digital Governance Capacity in Ecological Environments and Natural Resources is Steadily Improving

The fast collection of data on the Third Land Survey, land spatial planning, real estate registration, natural resource assets and natural geographical pattern form a data system with more than 5,000 image layers and more than 11 billion elements across the country. A three-dimensional map of natural resources and basic information platforms of land space have been basically built, and the functions of digital approval and supervision of natural resources have been successfully realized. The governance & application efficiency of meteorological big data has significantly improved. The meteorological big data cloud platform "Tianqing" (version 1.0) has been initially completed. Around 1,000 TB of annual meteorological data from 13 newly-added departments and localities (e.g., ecological environment department) plays a key supporting role in major fields and events like disaster prevention & mitigation, emergency rescue and air pollution prevention & control. The data of more than 600,000 water conservancy objects are updated on the map of the water conservancy department, which form a river & lake management data system that covers 800,000 river sections, 300,000 river heads above the town level and 600,000 river heads at the village level. Based on AI and remote sensing technology, information-extraction technology of "regulating illegal riverside occupation, construction, mining and waste" is widely used. The comprehensive sensing and monitoring capacities over forest & grass are continuously optimized, with the national forest & grass informatization rate reaching 81.74%.[18]

3. The Construction of a New-Type Smart City Is Furthered

In 2021, the construction of smart city clusters continued to advance. The integration of data, interface and standards among city clusters was deepened, with an emphasis on the coordination of government service areas. In 2019, the Yangtze River Delta explored the inter-city government services and took the lead in implementing "one-website handling" of government services. "Wu Gan Man You", an app for government services in the Yangtze River Delta, was launched, and the first-batch 51 items were handled in 14 cities. In 2020, the "Pan Pearl River Delta

[18] Data Source: *Digital China Development Report 2020.*

Inter-Provincial Service" Area was officially started on Guangdong Government Services Website, which can handle a total of 470 government services in Guangxi, Hainan, Hunan and Fujian in a one-stop manner. Citizens can enjoy government services of "handling affairs immediately, handling affairs online, handling affairs locally and handling affairs in one region". Meanwhile, City Information Modeling (CIM) has become a new direction in the construction of smart cities. In June 2020, the Ministry of Housing and Urban–Rural Development, in collaboration with the Ministry of Industry and Information Technology and the Office of the Central Cyberspace Affairs Commission, issued the *Guiding Opinions on the Construction of City Information Modeling (CIM) Basic Platform*, which required to "comprehensively promote the construction of CIM basic platform and the wide application of CIM basic platform in the field of urban planning, construction and management", and to "drive the application of autonomous and controllable technology and the development of related industries, and improve the level of urban optimized and intelligent management". Guangzhou, Xiamen, Beijing Municipal Administrative Center, Nanjing, Xiong'an New Area and China-Singapore Tianjin Eco-City start the construction of CIM infrastructure platform. To accelerate the construction of a new-type urban infrastructure that features digitalization, networking and intellectualization, in October 2020, the Ministry of Housing and Urban–Rural Development organized 16 cities (including Chongqing, Taiyuan and Nanjing) to take part in a "new urban construction" pilot, among which the construction of CIM infrastructure platform played an indispensable role.

4. The Construction of Digital Villages Is Steadily Implemented

In 2020, the goal of digital villages was fulfilled as scheduled. 13,700 4G base stations have been deployed in rural areas in China, with the number of satellite broadcast users in rural areas reaching 146 million. More than 98% of administrative villages in China have been connected to both optical-fiber networks and 4G network. The number of rural Internet users has reached 309 million, and the Internet penetration rate in rural areas has reached 55.9% (by the end of 2020). The total online retail sales of 832 national poverty-stricken counties have achieved 301.4 billion *yuan*, with an increase of 26% year on year.[19] In September 2020, seven departments, including the Office of the Central Cyberspace Affairs Commission, the Ministry of Agriculture and Rural Affairs and the National Development and Reform Commission, jointly carried out national digital village pilot projects and chose 117 counties (cities and districts) as the first-batch national digital village pilot areas. The pilot project mainly includes seven aspects, i.e., strengthening new-generation rural information infrastructure, overall planning and design of digital villages, exploring new business forms of the rural digital economy, forming new models of rural digital governance, improving information service systems related to agriculture, rural areas and rural residents, optimizing the integration and sharing mechanism of facilities and resources, and forging a mechanism for the sustainable

[19] *China Youth Daily*: "*Key Points of Digital Village Development in 2021* is Being Formulated". March 20, 2021, http://news.cyol.com/app/2021-03/20/content_18,970,312.htm.

development of digital villages. These provide practical experience for comprehensively promoting the construction of digital villages. In 2021, the No. 1 Central Document *Opinions of the CPC Central Committee and the State Council on Comprehensively Promoting Rural Revitalization and Accelerating Agricultural and Rural Modernization* proposed to "implement the digital village construction and development project", and to vigorously facilitate the digital and intelligent development of rural digital infrastructure, smart agriculture, rural public service and social governance as well as rural digital inclusive finance and rural e-commerce, in order to accelerate the digital transformation of agriculture and rural areas.

4.6.2 The Information Services that Benefit Citizens Are Continuously Optimized

1. The Informatization of Education Accelerates

The opening and sharing of educational resources deepen, and the national public service system of digital educational resources improves. 212 online platforms at various levels are connected, with the total number of application visits reaching 360 million times. The National E-Learning Resource Center actively develops a substantial number of the high-quality e-learning resource library. In 2020, more than 79,000 courses were collected in the library, with more than 378,000 types of various media resources, covering academic education, non-academic education and public media materials. By the end of 2020, the number of MOOC (massive open online courses) in China had increased to 34,000, and the number of learners had reached 540 million. The coverage of the online teaching model continues to expand. The English versions of "iCourse" and "XuetangX", higher education online teaching international platforms, are listed in the UNESCO Global Education Consortium, with the first batch of more than 500 English courses available.[20] The construction of new-type education infrastructure quickens. In July 2021, six departments, including the Ministry of Education, released the *Guiding Opinions on Promoting the Construction of New-Type Education Infrastructure and Building High-quality Education Supporting System*, which clarified key directions of the construction of new-type education infrastructure, including 20 items in six categories, i.e., new-type information network infrastructure, new-type platform system infrastructure, new-type digital resources infrastructure, new-type smart campus infrastructure, new-type innovative application infrastructure and new-type trusted and secure infrastructure.

2. "Internet + Healthcare" Further Advances

The informatization project of national health security and the construction of provincial-coordinated regional national health information platform steadily

---

[20] Data Source: *Digital China Development Report 2020*.

progress. The establishment of three major databases of population information, resident's electronic health files and electronic medical records, as well as the catalog-standard system of healthcare big data resources, quickens. The construction of national platform produces initial results. Various provinces (autonomous regions and municipalities directly under the Central Government) have initially established provincial-coordinated regional platforms. More than 7,000 public hospitals at or above the second level have been connected to regional, national health information platforms, more than 2,000 third-level hospitals have basically realized the exchange and sharing of medical-service information in hospitals, and 258 prefecture-level cities have actualized "one card for all medical services" for regional medical institutions. These robustly support cross-hospital, cross-region and cross-level medical services. So far, more than 1,100 Internet hospitals have been built nationwide, forming a trend of overall linkage. More than 7,700 hospitals above the second level have established appointment diagnosis & treatment system, which provides online services and solves the problems for patients who have to queue for a long time and have difficulty in taking medicine. Medical institutions at or above the second level generally provide time-based appointments, intelligent medical guiding & triage services and waiting reminding service. The appointment time is reduced to less than one hour, and the priority is given to reserving appointment number sources for grass-roots medical and health institutions in the medical consortium, so as to promote the first diagnosis and facilitate two-way transfer treatment at the grass-roots level.[21]

3. Digital Social Security Service Deepens and Expands

The Ministry of Human Resources and Social Security uses "one card for all social security services" to improve digital social security services. 550 million e-social security cards have been issued, covering 350 million (26%) card-holders. Services are available in all e-social-security-card regions and 437 channels, providing 40 national "one card for all social security services" to the public as well as more than 700 local services in various provinces (autonomous regions and municipalities directly under the Central Government). Presently, e-social security card provides login and identity-authentication services for the national government service platform and 24 provincial government service platforms. In 2020, e-social security card services exceeded eight billion times in total. Online services, e.g., unemployment registration, unemployment insurance application and social security relations transfer, continuously expand via the use of e-social security cards. Around 70% of prefectures and cities have achieved employment registration and job-hunting registration with social security cards, around 90% of prefectures and cities have included pension insurance and unemployment insurance in social security cards, and 100% of prefectures and cities have realized medical treatment and medicine purchase with social security cards. In the Yangtze River Delta, human resources and social security departments of three provinces and one city, working with the big data management

[21] Data Source: Data of the National Health Commission, by July 2021.

department, transportation department and cultural and tourism department, accelerate "one card for all social security services". Hainan, the Pearl River Delta, Beijing-Tianjin-Hebei and Chengdu & Chongqing regions actively explore the application of social security cards in taking buses, borrowing books and visiting parks, promote the comprehensive financial application of social security cards in the Guangdong-Hong Kong-Macao Greater Bay Area, and initially form the application-dimension of "one card for all social security services" for residents.[22]

4. The Inclusiveness of Digital Service Amplifies

IT services accelerate the elderly-friendly and barrier-free transformation, with mobile apps, application software, the authorized agency in applet, relative & friend agency and other functions launched successively. Some voice software provides AI voice recognition and AI voice-transcription technology services for groups with hearing impairment. Various regions in China organize public lectures and training to help the elderly improve their digital skills. For example, in July 2021, the Hubei Youth Volunteers Association guided the "blue vest" public welfare action to conduct elderly-friendly public welfare services for the elderly, such as the publicity of combating fraud, teaching with the use of smartphones, and the lecture of combating fraud. Some public service departments provide special channels for the elderly who do not have smartphones or cannot operate various software, and open "alternative windows" at the same time, so that the elderly who are "digitally disconnected" can enjoy high-quality public services. For example, "health code free" channels are promoted in passenger stations. "Digital support" campaigns for poor areas and special social groups are implemented successively. In 2020, the National E-Learning Resource Center provided more than 2,000 vocational skills and practical-technology micro courses in agriculture & forestry, animal husbandry & fishery and rural administration to poor areas. Information-based villages and households continue to deepen. The skill training of mobile phone applications for rural residents has been organized for five consecutive years, with a cumulative audience of more than 100 million. The R & D of ethnic language informatization software is steadily promoted, and the construction of the corpus is continuously improved. 3.29 million pieces of voice tagging data and more than 50,000 pieces of broadcast level recording corpus are added.

4.7 Digital Technology Assists Governments to Prevent and Control the COVID-19 Pandemic

In July 2020, at the 12th meeting of the Central Committee for Deepening Overall Reform, General Secretary Xi Jinping emphasized that we should encourage the use of digital technologies like big data, AI and cloud computing to better play a

---

[22] Data Source: Data of the Ministry of Human Resources and Social Security, by July 2021.

supporting rolc in pandemic monitoring & analysis, virus tracing, prevention & treatment and resource allocation. Currently, the COVID-19 pandemic occurs frequently in some countries, and China faces ever-growing pressure from "defending externally against importation, defending internally against rebound". In some regions or industries, digital technologies such as big data, AI and cloud computing are used in the prevention & control of the COVID-19 pandemic, economic and social development and regional comprehensive governance, which achieves remarkable progress.

4.7.1 Big Data Serves as the "Strongest Brain" for the Pandemic Prevention & Control

The CPC Central Committee and the State Council coordinate and advance the pandemic prevention & control and socio-economic development, and encourage various regions and departments to fully apply digital technology, improve the accuracy of information management of the pandemic, strengthen decision-making, optimize resource allocation and maintain the normal prevention & control of the pandemic. Big data becomes the "strongest brain" and plays an important part in combating the pandemic in various regions and departments.

At the end of 2020, Hubei Province built the big data application system for health emergency command and the COVID-19 pandemic, which integrated the data of 119 medical institutions, 23 commissions, offices and bureaus, 5,270 pharmacies, 2,565 schools and 13 key places in Wuhan, and implemented "data-based management in the city". With the help of 5G and AI technology, face recognition and infrared temperature measurement equipment were deployed in medical institutions, emergency vehicles and other places to monitor the febrile people. 13 categories of risk early warning indicators, including syndrome, goods and food, and dynamic monitoring of infectious diseases, were created to achieve intelligent early warning, multi-touch and data-based closed-loop management. On average, more than three million pieces of data were monitored every day and 956,000 people were monitored, which basically realized full scene and wide coverage.[23]

Sichuan promotes the prevention & control the COVID-19 pandemic with "one code and two systems". Supported by Sichuan Tianfu Health Code, Sichuan Nucleic Acid Testing Information System and Sichuan Immunization Planning Information Management System, Sichuan forges closed-loop management and accelerates the pandemic prevention & control. With big data, Sichuan adjusts the red, yellow and

[23] Xinhua News Agency: "Wuhan Built Health Emergency Command & Pandemic Big Data Application System". December 19, 2020, http://www.gov.cn/xinwen/2020-12/19/content_5,571,253.html.

green codes in a timely and accurate manner, and quickly obtains the vaccination information of citizens, so that citizens can spend less time with the help of big data.[24]

4.7.2 Big Data Serves as the "Best Tool" in Production and Life

Under the background of the normalization of the pandemic prevention & control, the "health code" becomes a must for Chinese people to travel. In December 2020, relevant state departments issued the *Notice on Further Promoting "Internet + Medical Care" and "Five Ones" Service Actions*, which clearly required various regions to implement mutual recognition and "one code pass" of "health code" across the country. In March 2021, the National Health Commission announced that China had basically realized "one code pass" with the health code. Simultaneously, China actively promotes the automatic integration of information such as nucleic acid testing, vaccination and travel to high & medium-risk areas with health codes via technological means. Some regions make innovations to strengthen the management and service of "health code" in the pandemic prevention & control.

After several stages of iterative development, Beijing's "Jian Kang Bao" becomes a necessary in the daily life of citizens. "Jian Kang Bao" mainly goes through three stages:

(1) The exploration stage. In March 2020, Beijing "Jian Kang Bao" was staged, which accelerated face comparison and health query and other functions.
(2) The expansion stage. In June 2020, the functions of "Jian Kang Bao" diversified. With newly-added "code scanning registration of personal information" and "visitor information register", "Jian Kang Bao" supports convenient register visits in various places. Meanwhile, many use cases, in combination with "Jian Kang Bao", realize the intelligent application of data. Some hospitals install intelligent "Jian Kang Bao" self-service query systems. Patients can complete temperature measurement, "Jian Kang Bao" state query and other functions when they show their ID cards or medical insurance cards. Some nucleic acid testing institutions finish rapid information registration, photographing and temperature measurement through "Jian Kang Bao", which improves the registration efficiency.
(3) The improvement stage. In November 2020, Beijing "Jian Kang Bao" launched the service function of "health-code-aided query for the elderly and children". In December 2020, "Jian Kang Bao" intelligent application terminal, or "Health Baby", was launched. The elderly can complete temperature measurements, visit

[24] www.thecover.cn: "An Interview with Sichuan Big Data Technology Service Center: 'One Code & Two Systems' Helps the Province's Pandemic Prevention & Control". June 7, 2021, https://www.163.com/dy/article/GBT8I9SR0514D3UH.html.

registration and health query by swiping the elderly cards for the disabled citizens. By the end of February 2021, more than 200 parks, hospitals and commercial supermarkets had installed "Jian Kang Bao", with a total of more than 800,000 inquiries.[25] Additionally, "Jian Kang Bao" launches "nucleic acid & vaccine services inquiry", which makes it convenient for citizens to check their personal nucleic acid testing or vaccination at any time. The function of "Jian Kang Bao" continues to be upgraded and improved.

Citizens in Guangdong can see their health codes by swiping their ID cards, which helps to bridge the digital divide. In December 2020, "Yue Sheng Shi", a WeChat applet for Guangdong government services, officially started the health and pandemic prevention verification system. With the system, citizens who cannot show their health codes or the elderly who don't have smartphones or cannot use smartphones, will quickly complete the "people-card-code three in one" check by showing their ID cards. The "Yue Sheng Shi" platform sets up a "special service area for respecting and caring for the elderly" in line with the requirement of an elderly-friendly design. The elderly can handle medical treatment, travel, pension qualification certification and other services with their fingertips anytime and anywhere online.[26]

[25] Beijing Municipal People's Government: "One Anniversary! Beijing 'Jian Kang Bao' AI Application Meets the Needs of Citizen's Life in Various Forms". March 1, 2021, http://www.beijing.gov.cn/ywdt/zwzt/yqfk/jkbsy/202103/t20210301_2295095.html.

[26] www.xinhuanet.com: "Guangdong Helps the Elderly Solve the 'Digital Divide' by Swiping the ID Cards and Showing the Health Codes". January 1, 2021, http://www.xinhuanet.com/local/2021-01/01/c_1126937556.htm.

Chapter 5
The Construction of Internet Content

5.1 Outline

Presently, the rapid development of the Internet spurs profound changes in the way of information communication. The Internet has evolved from a simple information storage & communication tool to a medium, platform and ecology. Besides, the production, innovation and dissemination of Internet content display new features and development trends. In particular, after the outbreak of the COVID-19 pandemic, public opinion on social events is expressed on the Internet, which increasingly becomes an important force that affects society, politics, economy, culture, economy and other fields.

Now, with the development of new media technology and innovative measures of media convergence, the construction of Internet content progresses steadily, with four highlights in development.

5.1.1 The Construction of Positive-Energy Contents is Continuously Strengthened to Create a Clear and Pure Cyberspace

In 2021, closely following the instructions of General Secretary Xi Jinping from important activities and speeches such as inspecting local regions, presiding over major conferences, attending important international conferences, and so on, various websites foregrounded the details, innovated expressive forms, and staged a number of excellent works in diverse form and vivid content, creating the strongest voice in cyberspace. Media platforms represented by central news websites enthusiastically implemented the publicity of major issues and the construction of positive energy content. Through micro videos, special events, multimedia interviews, and other forms, media platforms extensively publicized the "Centenary of the Communist Party of China", "China's complete victory in the fight against poverty", the start

Chinese Academy of Cyberspace Studies, *China Internet Development Report 2021*,
https://doi.org/10.1007/978-981-19-9311-4_5

of the "14th Five-Year Plan" and other important contents, augmented the voice of mainstream public opinion, spread positive energy on the Internet, told the stories of China well, and created a positive atmosphere in cyberspace.

5.1.2 *Government-Service New Media Highlight Both Contents and Services and the Consumption Model of Internet Products Matures*

In the prevention & control of the COVID-19 pandemic, great efforts are invested to the construction of Internet government services and new media systems for government services, so as to create a new media matrix for government services with diverse type, rich content and innovative form in content and to improve government services for the pandemic prevention & control in function. Additionally, with the maturity of IT-application capacity, Internet-culture industry and Internet user's consumption model, the scale of content industries like online literature and online video grows stably. On this basis, it forms a trend to explore IP-based operation, original development, customized operation and new-field expansion.

5.1.3 *Multiple Entities Participate in Comprehensive Cyberspace Governance and Antitrust Governance Becomes a Hot Topic*

The Civil Code of the People's Republic of China officially came into force in January 2021, which regulated the behaviors of various actors in cyberspace and provided clearer guidance and stronger guarantee for their participation in cyberspace governance. Chinese government speeds up the process of law-based cyberspace governance and continues to launch governance actions, with emphasis on combating unfair competition and monopolistic behavior in the Internet field. Multiple entities like enterprises, Internet users and industrial associations adhere to law-based website construction and access, and actively participate in the comprehensive governance of the Internet, creating a clear and pure cyberspace.

5.1.4 *Media Convergence Develops in a Deeper Way and Technological Innovation Boosts Industrial Development*

In China, the construction of Internet content makes significant progress and faces an era of reform and innovation in the content industry and media industry. New content technologies have a far-reaching impact on all aspects of media content production. In media convergence & development, mainstream media actively explore innovative ways of cross-field cooperation, deeply cooperate with Multi-Channel Network (MCN), and develop new business models. In content form, Internet media integrate big screens with small screens, and use technologies like AI, virtual reality and video recognition to help content production and achieve industrial innovation and development.

5.2 Steady Development: New Achievements in the Construction of Internet Content

Over the past year, various Internet entities voluntarily took part in the construction of positive energy content and continually created Internet cultural products. The construction of new media for government services in China steadily advanced, and the construction of Internet content came to a new stage.

5.2.1 Various Internet Entities Play an Active Role in the Construction of Positive Energy Contents

In 2021, various Internet entities zealously participated in the construction of Internet positive energy content and jointly forged an ecological environment for positive energy content, which further strengthened online mainstream public opinion.

1. Mainstream Media and Websites Build Reporting Matrices and Play a Major Role in Guiding Public Opinion

2021 saw the Centenary of the Communist Party of China. At the ceremony marking the Centenary of the Communist Party of China, General Secretary Xi Jinping delivered an important speech. He sincerely looked back on the glorious journey the Party traveled over 100 years of struggle, set a high value on the great achievements made by the Party in uniting and leading Chinese people over the past 100 years, solemnly announced building a moderately prosperous society in all respects, best captured the great founding spirit of the Party, and comprehensively summarized the "Nine Musts" that required to draw strength from history and forge ahead for a better future. He called on all members of the Communist Party of China to strive for greater glory for the Party and the people on the journey ahead. General Secretary Xi Jinping's important speech was also broadcast online simultaneously. Hundreds of millions of Internet users witnessed the historic moment when General Secretary Xi Jinping attended the ceremony marking the Centenary of the Communist Party of China and delivered an important speech. Major mainstream news media, websites and local media demonstrated their prowess and started an upsurge in online dissemination of celebrating the Centenary of the Communist Party of China since the beginning of 2021, with various popular products springing up. From June 22 to July 6, 2021, the number of relevant reports and click on Internet reached a record high. On July 1, 2021, top 50 microblog hot search lists were all topics in relation to the Centenary of the Communist Party of China, with *c.* one billion discussions.

The New Media Center of *People's Daily*, in conjunction with local media, launched a series of online activities under the theme that "The CPC bears its eternal great cause in mind, and the centenary only ushers in the prime of life." With visual representation, interactive guidance and audience-targeted communication, the center combined online display with offline experience to endow cyberspace with a surge of positive energy. It produced a series of short videos like *1921–2021*, which highlighted the great changes and historic achievements that had taken place in China under the leadership of the Party.

China Daily launched the Chinese-English bilingual video *Looking for Answers: An American Communist Explores China* for the Centenary of the Communist Party of China, which was disseminated more than 14 million times worldwide in 48 h. It carried out an interview under the theme of "A Date with China in 2021" to display a real China to the international community in an all-round way. In Xinjiang, nine foreign Internet celebrities from eight countries, as well as editors of central news websites, visited Urumqi, Kashgar and Hotan, vividly told the development story of Xinjiang, and presented the happy life of the people in Xinjiang.

2021 saw a landmark in the process of China's modernization and the start of the "14th Five-Year Plan". In 2021, online activities under various themes continued to stage and arouse wide attention. www.cctv.com and other media organized "China Positive Energy 2021: Five Hundred" Internet Excellent Works Selection, Exhibition and Broadcast, with a total of more than 125,000 applications. In line with major themes such as celebrating the Centenary of the Communist Party of China, poverty alleviation, the achievements in combating the pandemic, building a moderately prosperous society in all respects and the Olympic Games, a number of influential works, activities and models with Internet positive energy that were recognized by Internet users or the public were collected, selected and broadcast. A strong spiritual force was forged in cyberspace to embark on a new journey of the "14th Five-Year Plan". Central news websites and new media centers of central news institutions focused on Tokyo 2020 Olympic Games. They accurately grasped the personality and information-receiving habits of new-generation Internet users (e.g., the "post-90 s" and the "post-00 s"), took advantage of microblogs and other Internet platforms well-received among them, guided Internet hot topics, continuously strengthened the production and supply of micro-videos and short videos, conveyed rich emotional experiences with visual language, and firmly attracted their attention with close-up shots.

2. Network Platform Actively Guides Public Opinion and Builds New Bases for the Dissemination of Positive Energy

On the basis of traffic advantage, major Internet platforms voluntarily cooperate with mainstream media and relevant governmental departments, participate in the production and dissemination of positive energy content, carry forward the socialist core values, and furnish cyberspace with positive energy.

In combination with their communication advantages, major Internet platforms launched a series of special activities to guide users to actively participate in the creation of positive energy content. Kuaishou launched the special activity of "Celebrating the Centenary of the Communist Party of China" and invited www.qstheory.cn to join the platform, which set off a surge of theoretical study of "Internet +". The activity "I & China" was launched to demonstrate the great Chinese spirit and Chinese strength. Excellent Party members and users were assembled in the activity themed "embroidering a Party flag and revisiting the oath of joining the Party together" to show the struggle stories of grass-roots Party members. bilibili platform concentrated on such themes as rural education, Two Sessions, traditional culture and revisiting martyr's footprints, and staged a series of creative activities. TikTok designed the special section of "the Centenary of the Communist Party of China", which tremendously stimulated the enthusiasm of creators for relevant topics and created an online creative environment with positive energy.

3. Individual Content Creators Play an Active Role in Promoting the Dissemination of Positive Energy Content

Under the guidance of mainstream media and platforms, more and more individual content creators voluntarily participate in the production and dissemination of positive energy content, which builds a nationwide content production & consumption system with wide coverage, high quality and deep influence via bottom-top linkage.

Presently, various short-video platforms have posted topics on "the Centenary of the Communist Party of China". By July 2021, Internet users had released more than 400,000 videos in this regard on TikTok, with a total viewing of more than 4.3 billion times. Creators on bilibili spontaneously created tags of "the Centenary of the Communist Party of China" and "positive energy" and paid respect to the Centenary of the Communist Party of China by various means of MV, hand-drawn work and Vlog. In May 2021, the China Internet Development Foundation, www.cctv.com, PING AN and Kuaishou jointly launched "Relay for Rural Revitalization". In less than one month, 641 videos were posted on the Kuaishou platform concerning the topic "I Join the Relay for Rural Revitalization", with a total viewing of 190 million times. Individual content creators played an active role in disseminating and popularizing positive energy content in cyberspace.

5.2.2 Internet Culture Services and Products Are Iteratively Innovated

In the past year, various types of Internet culture services & products developed steadily and kept pace with the times. With the continuous progress of IT and the improvement of technological-application capacity, the current Internet content products not only keep pace with the times in R & D and operation, but also stress user-oriented development, so as to quicken the iterative innovation of products.

1. The Ecology of Teleworking Application Further Improves "New Employment Form"

"New Employment Form" signifies a platform-based and employer-free flexible employment model that directly connects the two ends of supply and consumption via the Internet platform. "Supporting and standardizing the development of new employment form" has been included in the *Outline of the 14th Five-Year Plan for National Economic and Social Development* as an employment-priority policy.

By June 2021, the number of teleworking users in China had reached 381 million, with an increase of 35.06 million over December 2020, accounting for 37.7% of all Internet users.[1] The scale of teleworking application market maintained rapid growth. As the construction of new-type infrastructure (e.g., 5G network, big data center, AI and industrial Internet) continues to accelerate in China, the hardware & software basic capabilities of business-level applications remarkably improve, and key capabilities like massive data processing, cloud storage and cloud computing further increase. The functions of online meetings, online document collaborative editing, online task management and corporate cloud are re-organized, which forms a trend of ecology-oriented development of teleworking applications. Simultaneously, teleworking applications can break geographical restrictions, enrich the collaboration model of organizational members, reconstruct the organizational operation process, provide important digital support for new digital-technology-based jobs like online delivery workers, promote the development of "new employment forms", and boost digital transformation and upgrading.

2. Online Education Lays Stress on Quality, with Industrial Rectification Promoted in an Orderly Manner

With progress made in the prevention & control of COVID-19, universities and primary & secondary schools basically resumed normal teaching order, and the scale of online education users declines. By June 2021, the scale of online education user in China had reached 325 million, with a decrease of 16.78 million over December 2020.[2] Online education comes to the development stage that emphasizes the qualitative change of content and services.

On the one hand, online education platforms launch public courses. Several organizations provide online livestreaming courses that cover main subjects in primary, middle and high schools for many regions for free. On the other hand, many online education brands use AI, big data and other technologies to assist teachers in personalized teaching and provide intelligent teaching services for users. For example, some

---

[1] Data Source: *The 48th Statistical Report on China's Internet Development* released by the China Internet Network Information Center.

[2] Data Source: *The 48th Statistical Report on China's Internet Development* released by the China Internet Network Information Center.

online education platforms adopt the "intelligence + interaction" model to complete the process of teaching evaluation and accumulate learning data via interactive feedback, so that users can obtain more targeted teaching content in the next stage of learning.

Meanwhile, regulatory authorities further strengthen the standardization and rectification of the online education and training industry. In November 2020, the Cyberspace Administration of China and the Ministry of Education issued the *Notice on Further Strengthening the Standardized Management of Online Lesson Platforms Involving the Minors.* The *Notice* advocates that "local Internet information and education departments should actively cooperate with industrial associations, social organizations, primary & secondary schools and parent's committees to regularly evaluate the governance effectiveness of online lesson platforms involving the minors". It states that "the society and the public can report relevant illegal clues to local Internet information departments", mobilizing all sectors of society to participate in the regulation and supervision of online lesson platforms.[3] In May 2021, the State Administration for Market Regulation imposed a warning and a top fine of 2.5 million *yuan* on two off-campus training institutions respectively, for conducting misleading business publicity or using false or misleading price means to lure consumers into trading. In June 2021, the State Administration for Market Regulation imposed top fines on 13 off-campus training institutions for false publicity and price fraud. In June 2021, the Ministry of Education established the Department of Supervision for Off-Campus Education and Training to further normalize and improve the supervision and regulation of the online education and training industry. Regulatory measures are effectively implemented, which further promotes the standardization and order of the online education and training industry and better accelerates the development of the industry. In July 2021, the General Office of the CPC Central Committee and the General Office of the State Council issued the *Opinions on Further Easing the Burden of Homework and Off-campus Training on Students in Compulsory Education Stage*, which proposed to actively explore the use of AI technology to reasonably control the continuous online training time of students and forbade online training institutions not to provide and disseminate "solving problems with photo search" and other bad learning methods that would make students lazy to think, affect student's independent thinking and violate the laws of education and teaching.

3. The Online Audio Industry Grows Steadily and Platform Enterprises Deploy IoV Ecology

Now, the online audio industry in China grows stably in the scale of users, mostly young users. As relevant data suggested, the number of online audio users in China

---

[3] www.cac.gov.cn: *The Notice on Further Strengthening the Standardized Management of Online Lesson Platforms Involving the Minors.* December 3, 2020, http://www.cac.gov.cn/2020-12/03/c_1,608,562,096,397,737.html.

reached 570 million in 2020 and would reach 690 million in 2022, as estimated, most of whom were young people, with those under 30 years old accounting for 66.7%.[4]

Simultaneously, the wide IoV market arouses the attention of audio platforms. In September 2020, Dragonfly FM and Baidu Apollo IoV agreed to cooperate in a car audio applet, which combined Baidu's big data recommendation to provide content services for users. In November 2020, Ethereal Sound launched by the China Media Group joined BYD's app market and Huawei's car infotainment products, which started the layout of IoV business and further expanded the use cases of online audio. In December 2020, Li Zhi, a professional audio sharing platform, announced its cooperation with XPENG in the direction of car audio, and launched relevant car audio products in the car app market of XPENG.

4. Online Literature Becomes Important Cultural Resources and IP-Based Exploration Enters a New Stage

In recent years, online literature aroused widespread attention and evolved into important cultural resources. By December 2020, the scale of online literature users in China had reached 560 million, with an increase of 4.75 million over March 2020, accounting for 46.5% of all Internet users.[5]

Additionally, the development of online literature stirs the prosperity of the IP-adaptation market, and literary works enter the stage of IP-based development. In March 2021, the Chinese Academy of Social Sciences released the *2020 China Online Literature Development Report*, which confirmed that with the modern industrial layout basically completed in China's cultural industry, IP cultural-production thinking like "literature-comics-film-game" linkage became the consensus for IP development of online literature industry, and that China's cultural industry came to new development stage. In 2020, online literature platforms represented by China Literature connected the operation & development forms of IP at various stages and realized high-quality and efficient IP adaptation of online literature, by integrating resources of Tencent Pictures, New Classics Media and China Literature Film & TV and launching "300 Online Literature & Comics Adaption Program".[6]

5. Mobile Games Develop Rapidly and Cloud Games Become Hot Industrial Field

In 2021, the online game industry in China maintained momentum for development. Relevant data at the end of 2020 disclosed that the actual sales revenue of China's online game market realized 278.686 billion *yuan*, with an increase of 47.81 billion *yuan* over 2019, up by 20.71% year on year.[7] Specifically, mobile games maintained

[4] www.iimedia.cn: *Research Report on China's Online Audio Industry 2020–2021*. March 31, 2021, https://www.iimedia.cn/c400/77771.html.

[5] Data Source: *The 47th Statistical Report on China's Internet Development* released by the China Internet Network Information Center.

[6] www.cssn.cn/: *Report on the Development of China's Internet Literature in 2020*. March 18, 2021, http://www.cssn.cn/wx/wx_xszx/202103/t20210318_5,319,695.shtml.

[7] Data Source: *Report on China's Game Industry in 2020* released by China Audio–Video and Digital Publishing Association Game Working Committee and China Game Industry Research Institute.

a high share and business revenue in China's online game market and took the leading position. In 2020, the actual sales revenue of China's mobile game market reached 209.676 billion *yuan*, with an increase of 51.565 billion *yuan* over 2019, up by 32.61% year on year, accounting for 75.24% of the overall revenue of the game market. In January 2021, the National Press and Publication Administration published the approval information of the first-batch domestic online games in 2021, and a total of 84 domestic mobile games obtained publication numbers through approval.[8] The thriving online game market created new opportunities for originally-developed games. In 2020, the growth rate of sales revenue of Chinese independently-developed online games in overseas markets increased again. The actual sales revenue in overseas markets reached 15.45 billion U. S. dollars, with a year-on-year increase of 33.25% as well as a trend of fast growth. The sales revenue in overseas markets exceeded 100 billion *yuan* for the first time.[9]

Besides, with the aid of technological development, the scale of cloud gaming continues to grow. In 2020, the scale of cloud gaming in China achieved around 1.2 billion *yuan*, doubling year-on-year. By 2022, the market scale would reach 4 billion *yuan* as estimated. Meeting the needs of fragmentation, cloud gaming help to expand new business models like timed payment and member subscription, and will develop rapidly. Hopefully, the development of native cloud gaming content and engines establishes a technological foundation for the construction of "super digital scene" in games.[10]

6. With the Innovation of Video Form, Mini-series has Become a New Field

The short video market in China develops fast and makes continuous progress. By June 2021, the scale of short video users in China had reached 888 million, with an increase of 17.07 million over December 2020, accounting for 87.8% of all Internet users.[11] With the popularity of the short video market, major platforms accelerate the layout of an innovative form of drama like short dramas and vertical-screen dramas and vie for the dominant position in competition. In October 2020, Kuaishou, a short video platform, introduced an excellent mini-series. By December 2020, "Kuaishou Little Theatre", a mini-series channel, had collected more than 20,000 mini-series, of which more than 2,500 had been played more than 100 million times.[12]

---

[8] National Press and Publication Administration: "The Approval Information of Domestic Online Games in January 2021". January 27, 2021, http://www.nppa.gov.cn/nppa/contents/320/75551.shtml.

[9] Xinhua News Agency: "The Overseas Revenue of Domestic Games Will Exceed 100 Billion *Yuan* for the First Time in 2020". December 20, 2020, http://www.xinhuanet.com/enterprise/2020-12/20/c_1,126,883,870.htm.

[10] *The Development Report of China's Internet Copyright Industry 2020* released by the National Copyright Administration Research Base of the Internet Copyright Industry.

[11] Data Source: *The 48th Statistical Report on China's Internet Development* released by the China Internet Network Information Center.

[12] *Beijing Business Today*: "Short, Meaningful and Interesting: Kuaishou and Wesee See the Rise of Mini-series". December 30, 2020, https://baijiahao.baidu.com/s?id=1687493096335943876&wfr=spider&for=pc.

In December 2020, Tencent Wesee announced to officially launch "Mars", a well-produced vertical-screen continuous mini-series brand with a playing time of 1–3 min, started the support plan for high-quality mini-series "Mars Plan", and jointly promoted the development of vertical-screen IP mini-series industry with creators.[13] In June 2021, the TikTok platform organized the "Mini-series is Interesting · 2021 Mini-series Launching Ceremony" and started the "Mini-series New Season Plan Version 2.0", in which a total of 1,400 creators participated.[14] On the basis of resource advantages, major video platforms make a foray into new mini-series markets and propel platforms and individual creators into building the mini-series ecology.

5.2.3 The Interactivity of Communication Products on Chinese Culture is Enhanced

Over the past year, China confidently responded to the international public opinion environment, actively used various forms of new media, launched highly-interactive cultural communication products, conveyed the true voice of China to the world, spread China's stories, displayed the achievements of China's development, and vigorously shaped a trustworthy, energetic and respectable image of China.

1. "Going Global": Promoting the Overseas Dissemination of Chinese Cultural Products

On the eve of complete victory in the fight against poverty, the great success of poverty alleviation in China has been extensively disseminated overseas via various media. Chinese media innovate the reporting model and apply short videos, animation, data dynamic charts and other forms to vividly depict a documentary picture of China's poverty alleviation. In December 2020, China.org.cn staged a series of bilingual short videos of *Statistical Data of China's Poverty Alleviation.* In the form of animation and ink-wash painting, the videos demonstrated great changes in China in an all-round way with a deep Chinese style. More than 100 overseas media, including the Associated Press News and Yahoo News, re-transmitted the videos for many times. *China's Rural Vitalization* invited a number of international celebrities in relevant fields online to write a series of comments, and some articles were reprinted by more than 400 foreign media like the websites of the Associated Press and NASDAQ.[15]

Additionally, Chinese cultural activities actively go global, which triggers heated discussions in different countries and regions. In April 2021, the "Hebei Cultural

[13] *Shenzhen Economic Daily*: "With One Billion *Yuan* Invested to Mini-series, Tencent Wesee Supports the Development of Vertical-Screen IP Short Drama Industry". December 23, 2020, https://xw.qq.com/cmsid/20201223A0DLFD00.

[14] www.ynet.com: "TikTok Held a Release Conference of 'Mini-series is Interesting', Whose New Season 2.0 Supports Mini-series Creators". June 11, 2021, http://ent.ynet.com/2021/06/11/3317227t1254.html.

[15] Jiao Meng and Wu Qiongjing: "Tell a Good Story About China's Poverty Alleviation via Media Convergence". *International Communications*, 2021 (01): 47–49.

Tourism Month" Online Promotion came to Morocco and organized exhibitions on social platforms for five weeks. Scenic spots in Hebei, like Chengde Mountain Resort, Shanhai Pass Great Wall, and Bai-yang Lake, as well as Hebei folk arts like Wuqiang New Year Painting, Letting Shadow Play, Xinji Peasant Painting and Wu'an Nuo Opera, were highly praised by Moroccan people. Morocco Mohammed V National Theatre forwarded the exquisite acrobatic repertoire presented by Wuqiao Acrobatic Troupe on its official Facebook account.[16]

2. "Introducing Overseas Creators": Encouraging Overseas Content Creators to Tell China's Stories

In the new era, China actively explores various forms of international communication. While improving the content-construction mechanism, China welcomes overseas content creators to come to China and tell real China's stories together.

In April 2021, the State Council Information Office launched "The Centenary of the Communist Party of China & The Journey to Old Revolutionary Bases" overseas media interview activity. More than 20 foreign correspondents in China from Reuters, Agence France-Presse, *The Wall Street Journal* and other media re-visited the 100-year development path of the Communist Party of China, deeply reported the industrial development of Jiangxi, Guizhou, Shaanxi and Hebei, understood the struggle history of the Communist Party of China, and felt the spiritual outlook of the Communist Party of China.

Additionally, China gives full play to the non-governmental model in international communication, tells the development story in the most down-to-earth way, and shapes a trustworthy, energetic and respectable image of China, with the strong communication power of We Media and online celebrities. In April 2021, the Ministry of Foreign Affairs hosted "'Heroic Hubei: Reborn for New Glories': Presenting Hubei Province to the World—Special Event of the Ministry of Foreign Affairs" both online and offline, which displayed the achievements of Hubei's combating the pandemic and presented a new look of Hubei's economic and social development in an all-round way. In order to coordinate with the special promotional event, Hubei launched a large-scale multi-media interview of "Decoding Happy Wuhan by Foreign Online Celebrities". Overseas foreign online celebrities who worked and lived in China for a long time were invited to come to Wuhan together with the multilingual anchors of the China Media Group to witness the development of Wuhan today. Simultaneously, video & social platforms at home and abroad, such as Vlog, Facebook, YouTube, bilibili and Sina Weibo, disseminated the development stories of China.

[16] Xinhua News Agency: "'Hebei Cultural Tourism Month' Was Well Received in Morocco". April 30, 2020, https://baijiahao.baidu.com/s?id=1,698,474,602,495,192,783&wfr=spider&for=pc.

5.2.4 *Government-Service New Media Highlight the Nature of Service*

In 2021, to prevent and control the COVID-19 pandemic, the construction of China Internet government services & government service's new media system continued to improve and made important contributions to effectively curbing the spread of the pandemic.

By December 2020, the scale of Internet government service users in China had reached 843 million, with an increase of 21.6% over March 2020, accounting for 85.3% of all Internet users.[17] Various regions and departments vigorously promote "one-website handling", and the average commitment time limit for more than 50% of administrative licensing matters in China diminished by more than 40%. In the COVID-19 pandemic, the National Integrated Government Service Platform launched the "pandemic prevention health code", which supported "one code pass" in most regions and signaled an important innovation in big-data-backed pandemic prevention & control, with around 900 million times of applications and more than 40 billion of users. According to the *UN E-Government Survey 2020*, the ranking of China's e-government development index has risen to 45th in the world, 20 places higher than the ranking in 2018. In particular, the ranking of the online service index jumped from the 34th to the 9th in the world, leading the list in the world.

Simultaneously, various regions and departments actively build new-media-service matrices for government services with diverse types, rich contents and innovative forms. By December 2020, governments at various levels had opened a total of 82,958 governmental headlines and 26,098 governmental accounts on TikTok. There were 140,837 Weibo accounts of governmental agencies certified by the Sina platform, which covered 31 provinces (autonomous regions and municipalities directly under the Central Government). The construction of county-level media-convergence centers has been systematically carried out nationwide, with a blowout in the number.[18]

Governmental new media at all levels give full play to the advantage of the matrix, actively undertake the functions of authoritative information release and government service, combine local characteristics and strengths, and produce a large number of high-quality and interesting new media products, which effectively forge a force in guiding mainstream public opinion. In April 2021, the announcement ceremony of the 2020 Outstanding Media-Convergence Works Collection and Selection hosted by www.gmw.cn and other organizations was staged online and offline simultaneously. Ten new media works produced by media-convergence centers of regions or departments (e.g., Beijing, Guangdong, Henan, Jiangsu, Shandong, Shanxi and Sichuan),

---

[17] Data Source: *The 47th Statistical Report on China's Internet Development* released by the China Internet Network Information Center.

[18] Data Source: *The 47th Statistical Report on China's Internet Development* released by the China Internet Network Information Center.

including *The Civilized Bus Tour of the Capital*, *ISZ* and *Gaochun Crab in Rap*, were rated as excellent cases of district & county media-convergence centers in 2020.[19]

5.2.5 The Influence of Public Communication Enlarges

With unique advantages and innovative potential in information communication, the Internet diversifies the communication forms of public activities, enhances the accessibility of diverse audience groups, and helps to enlarge the influence of public communication.

1. The Popularity of Public Communication: Internet Enterprises Boost the Diversification of Public Communication

As the *2020 Corporate Public Communication Report* indicates, the communication paths of public projects of Internet enterprises are increasingly diversified, with an eye-catching performance of public projects in various forms.[20] By dint of traffic advantages of their platforms, Internet public platforms operated by Internet enterprises popularize public communication and expand charitable donation cases. The COVID-19 pandemic clogs the circulation and sales of agricultural products. To solve this problem, Pinduoduo launches "combating the pandemic and helping farmers" and adopts new e-commerce-consumption poverty-alleviation models, in which city and county heads serve as network anchors and farmers sell more goods. Tencent "99 Giving Day" invests funds and resources to help public charities in public fundraising, information disclosure and brand construction in a high-quality way, guide & encourage public charities to connect with a wider and more active population, caring enterprises and social organizations, explore more diverse and effective charitable donation cases, and build a healthier and more sustainable Internet charity ecology.

2. Short Video: New Trend of Public Communication

During the COVID-19 pandemic, short video & livestreaming industry develops rapidly. In the duration, short video user comes second to socializing app. Short video and livestreaming become new choices for the content dissemination of public organizations and promote the channel of content down to larger social groups. Tencent Charity works with public organizations like the SEE Foundation as well as the IP of public communication platforms like Zhihu. The livestreaming at rarely-visited nature reserves evokes public attention and enables the public to understand the details of public projects on mobile screens.

---

[19] www.gmw.cn: "Representative Cases of '2020 Outstanding Media-Convergence Works Collection and Selection' Were Announced". April 16, 2021, https://topics.gmw.cn/2021-04/16/content_3 4,768,262.htm.

[20] www.newsgd.com: "Public Welfare Communication Popularizes Comprehensive Poverty Alleviation Model". December 16, 2020, http://epaper.southcn.com/nfdaily/html/2020-12/16/con tent_7,919,870.htm.

5.3 Joint Contribution: New Ideas for Comprehensive Cyberspace Governance

The construction of Internet content plays a key role in purifying the cyberspace environment and improving the national comprehensive cyberspace governance system. In the past year, with the joint efforts of governments, enterprises, the public and industrial associations, remarkable achievements were made in comprehensive cyberspace governance. Besides, Internet-platform monopoly increasingly intensifies, seriously disturbs market order, and causes many problems in the Internet-content ecological environment. In February 2021, the Antitrust Commission of the State Council issued the *Antitrust Guidelines for Platform Economy*, which provided ideas for the anti-monopoly of Internet platforms and the governance of disorderly expansion of capital.

5.3.1 Promoting Law-Based Cyberspace Governance and Valuing the Security and Development of Internet Content

To better promote the construction of cyberspace ecology, since 2020, relevant departments in China have focused on controlling the chaos of online livestreaming, regulating the collection of personal information on Apps and standardizing public accounts in their administrative and legislative work. Regulatory documents have been successively released, including the *Regulations on the Management of Public Account Information Services for Internet Users (Revised), Guiding Opinions on Strengthening the Standardization and Management of Online Livestreaming, Regulations on the Scope of Necessary Personal Information for Common Types of Mobile Internet Applications*, and *Measures for the Management of Online Livestreaming Marketing (For Trial Implementation). The Civil Code of the People's Republic of China* officially came into force in January 2021, which regulated data, Internet virtual assets, electronic contracts, personal-information protection and Internet tort liabilities, standardized the behaviors of various entities in cyberspace, and ameliorated a comprehensive cyberspace governance system.

In law enforcement, the Cyberspace Administration of China has organized and launched special rectification actions for many times. For example, in November 2020, the Cyberspace Administration of China intensively rectified the problems of "paid deletion of posts" and "soft porn" in cyberspace,[21] and supervised and inspected

[21] www.cac.gov.cn: "The Cyberspace Administration of China Intensively Rectifies the Problems of 'Paid Deletion of Posts' and 'soft porn' in cyberspace". November 5, 2020, http://www.cac.gov.cn/2020-11/05/c_1606139352605590.htm.

the *pêle-mêle* of information content on mobile apps.[22] The Cyberspace Administration of China started the "Nationwide Campaign to Purify Online Environment · The Spring Festival" in February 2021,[23] "Nationwide Campaign to Purify Online Environment · Undesirable Fan Culture" (two months) in June 2021,[24] and initiated the "Nationwide Campaign to Purify Online Environment · Commercial Websites & Platforms" as well as combating illegal collection, compilation and release of financial & economic information on We Media in August.[25] These aimed to standardize the order of Internet communication and purify the environment in cyberspace. Intensive measures are taken in the legislation and law enforcement of Internet governance, which mirrors the systematization and normalization of Internet content governance.

5.3.2 Corporate Entities Take More Initiative and Establish Industrial Standards

Internet enterprises take the initiative in the construction and governance of Internet content, respond positively to the government's administrative guidance and initiatives on law & compliance-based operation, and collaborate with social organizations on industrial self-discipline in cyberspace.

1. Focusing on Corporate Social Responsibility, Internet Enterprises Contribute to Poverty Alleviation

As their social influence increasingly enlarges, Internet enterprises actively undertake social responsibilities and make a contribution to poverty alleviation in China. In the "the last mile" of poverty alleviation, major Internet enterprises continue to deepen their efforts in rural e-commerce poverty alleviation, rural education poverty alleviation and other fields, and use the functions of platforms (e.g., information production & distribution, data collection & processing, etc.) to benefit the vast

---

[22] www.cac.gov.cn: " The Cyberspace Administration of China Rectifies the *Pêle-Mêle* of Information Content on Mobile Apps". November 5, 2020, http://www.cac.gov.cn/2020-11/05/c_1606140440558135.htm.

[23] www.cac.gov.cn: " The Cyberspace Administration of China started '2021 Nationwide Campaign to Purify Online Environment · The Spring Festival'". February 4, 2021. http://www.cac.gov.cn/2021-02/03/c_1613923672042416.htm.

[24] www.cac.gov.cn: "Office of the Central Cyberspace Affairs Commission Started 'Nationwide Campaign to Purify Online Environment · Undesirable Fan Culture'". June 15, 2021, http://www.cac.gov.cn/2021-06/08/c_1624735580427196.htm.

[25] www.cac.gov.cn: "The Cyberspace Administration of China Started 'Nationwide Campaign to Purify Online Environment· Special Rectification of Illegal Collection, Compilation and Release of Financial Information by Commercial Website Platforms and We Media'". August 27, 2021, http://www.cac.gov.cn/2021-08/27/c_1631652531513374.htm.

population in poor areas.[26] Technologies like 5G and AI are closely integrated with corporate public projects, which gives full play to the IT advantage of enterprises.

2. Reinforcing Industrial Exchanges, Internet Enterprises Enhance Industrial Self-Discipline

Internet enterprises discuss the topic of social responsibility in the form of a forum and achieve industrial standardization and self-discipline. In December 2020, the 7th China Internet Corporate Social Responsibility Forum was staged in Beijing under the theme of "Responsibility Guides and Forges New Patterns of Industrial Development". At the forum, participants shared corporate-social-responsibility reports and deeply exchanged their experiences and achievements in serving society, and produced admirable results. In order to better realize corporate communication and standardize self-discipline, relevant association standards are constantly implemented. For instance, at the end of 2020, association standards, such as the *Guidelines for the Compilation of Internet Corporate Social Responsibility Report*, *Framework for the Construction of Harmful Information Prevention and Control System for the Minors by Internet Operators*, *Guidelines for Artificial Intelligence Data Labeling for Content Security*, and *Basic Requirements for AI-Based Multi-media Content Identification*, were released, which can help enterprises to strengthen communication with various stakeholders and better achieve the goal of comprehensive governance.

3. Responding to the Initiatives of Governments and Industrial Associations, Internet Enterprises Actively Fulfill Corporate Commitments

Internet enterprises energetically respond to the call of industrial associations. In December 2020, under the guidance of the Cyberspace Administration of China, the National Development and Reform Commission, the Ministry of Commerce, the State Administration for Market Regulation and Shandong Provincial People's Government, and organized by the China Federation of Internet Societies, the 2020 China Internet Integrity Conference was successfully held at Qufu, Shandong, under the theme of "Trustworthiness & Mutual Trust in Practice: Jointly Promoting the Construction of Internet Integrity". The conference built an exchange platform for experience in the construction Internet integrity, strengthened the self-discipline of the Internet industry, boosted the construction of Internet integrity, advocated Internet civilization, and quickened the construction of a comprehensive Internet governance system and the development of the digital economy. In April 2021, the first batch of 150 enterprises signed the *Initiative for Internet Platform Enterprises to Operate in Accordance with Laws and Regulations*, promising to firmly follow the right direction, enhance the awareness of responsibility, uphold the priority of national interests, operate in line with laws and regulations, fulfill social responsibilities, and urge platform economy to serve high-quality development and high-quality life. In July 2021, the 2021 China Internet Integrity Conference was successfully launched

[26] www.bytedance.com: *2020 ByteDance Corporate Social Responsibility Report*, December 18, 2020, https://lf3-static.bytednsdoc.com/obj/eden-cn/uj_shpjpmmv_ljuhklafi/ljhwZthlaukjlkulzlp/csr/csr-2020.pdf.

in Changsha, Hunan, under the theme of "Sincerity and Integrity as a Force for Development: Jointly Building Internet Integrity in the New Era". More than 500 industrial guests shared their ideas on Internet integrity in the new era and promoted the joint contribution and shared benefits of Internet-integrity space.

5.3.3 Social Supervision on Internet Content Becomes Prominent and Pools Efforts in Internet Governance

In the past year, all sectors of society in China played a greater role in participating in the supervision of Internet content. Major websites continuously opened, improved and publicized new and old reporting channels, and Internet commercial platforms voluntarily optimized reporting work, which created conditions for the public to enhance their sense of online reporting.

In 2020, online reporting departments at various levels accepted 163 million reports, with a year-on-year increase of 17.4%. Specifically, the Office of the Central Cyberspace Affairs Commission Illegal and Harmful Information Reporting Center accepted 2.288 million reports, down by 5.2% year on year. Local Internet information departments accepted 15.962 million reports, down by 27.1% year on year. Major websites accepted 145 million reports, with a year-on-year increase of 26.1%, of which major commercial websites like Weibo, Baidu, Alibaba, Tencent, Sina and Toutiao accounted for 79.6%, reaching 115 million. In November 2020, the Office of the Central Cyberspace Affairs Commission Illegal and Harmful Information Reporting Center organized the seventh batch of 500 websites to uniformly publish the reporting and acceptance methods. More than 3,100 websites have announced online supervision channels to the public, which not only enables Internet users to master more channels for website reporting, but also facilitates websites to finish online reporting work and effectively fulfill the primary responsibility.

Refuting the rumor also constitutes an important part in Internet content governance. The foundation and mechanism of online rumor governance in China continuously improve. The joint rumor refutation mechanism led by the China Internet Joint Rumor Refutation Platform and co-launched by central media platforms and provincial rumor refutation platforms plays a prominent role. For example, the China Internet Joint Rumor Refutation Platform, in conjunction with "Xue Xi Qiang Guo", launched the "rumor refutation platform" channel in March 2021. Consisting of five parts, i.e., "governance dynamics", "exposure platform", "lecture hall", "caution," and "report verification", it publicizes policies and regulations on cybersecurity and Internet rumor governance, strengthens the public ability to distinguish rumors and false information, and significantly improves the influence and authority of online rumor refutation.

The 2020 China Internet Rumor Refutation Excellent Works Selection was hosted by the China Internet Joint Rumor Refutation Platform, under the guidance of the

Office of the Central Cyberspace Affairs Commission Illegal and Harmful Information Reporting Center. In January 2021, 20 image-text rumor-refutation works and ten creative rumor-refutation short videos won awards. As an important part of online rumor governance, the activity has been successfully organized for three sessions since 2018, with wide social influence.

In April 2021, in response to rumors about Xinjiang cotton started by some foreign enterprises, the Internet Information Office of Xinjiang Uygur Autonomous Region launched the Xinjiang Internet Joint Rumor Refutation Platform. Simultaneously, the China Internet Joint Rumor Refutation Platform forwarded relevant topics on the microblog. Foreign envoys and diplomats in China, experts in relevant fields and ordinary Xinjiang Internet users extensively refuted rumors about Xinjiang, actively corrected misunderstandings and told the truth from various angles.

Relevant governmental departments attach great importance to the role of social collaborative governance while launching special campaigns and issuing administrative documents. For example, since November 2020, the Cyberspace Administration of China has engaged in special rectification against the chaos of mobile app information contents, encouraging the vast number of Internet users to actively provide clues and verify the reported clues in line with law and regulations.

5.3.4 The Digital Literacy of Internet Users Continuously Improves and Their Enthusiasm Runs High

In 2021, the Cyberspace Administration of China, in collaboration with the Ministry of Education, the People's Bank of China, the All-China Federation of Trade Unions, the Central Committee of the Communist Youth League, All-China Women's Federation and other ministries and commissions, deeply organized or implemented the project of "making good Internet users in China", extensively conducted special activities of "making good Internet users in school", "making good Internet users in finance", "making good Internet users among workers", "making good Internet users among the youth" and "making good Internet users among female citizens", and vigorously promoted civilized Internet access and civilized Internet use in the whole society. In March 2021, the Office of Hubei Cyberspace Affairs Commission and the School of Journalism and Communication, Wuhan University jointly launched the "Internet Literacy Community Class", which provided Internet literacy classes to the elderly in communities and enabled the elderly with difficulties in using the Internet to experience the convenience of digital life.[27] Besides, in order to enhance the richness and diversity of Internet literacy education projects for the youth in schools, education departments in various places jointly launched a series of activities of "Internet civilization literacy at schools" with local primary & secondary schools, and promoted the construction of Internet literacy education system that connected

[27] *Hubei Daily*: "Internet Literacy Community Classroom Helps the Elderly Adapt to New Internet Life". March 23, 2021, http://www.cac.gov.cn/2021-03/23/c_1618075676626187.htm.

families, schools and society, by taking forms loved by the youth like holding exhibitions offline, offering courses online, and distributing Internet literacy books and publicity materials[28,29,30].

5.3.5 *Emphasis is Placed on Internet Antitrust Governance and the Prevention of Disorderly Expansion of Capital*

The Internet industry penetrates into all aspects of social life. Presently, the monopoly of super Internet platforms has intensified. Internet enterprises that occupy the monopolistic position in the market greatly affect the distribution and presentation of Internet content and become the key field of comprehensive Internet governance.

At the end of 2020, anti-unfair competition and antitrust supervision aroused wide attention in Internet governance. Several central meetings stress the necessity to "strengthen anti-monopoly and prevent the disorderly expansion of capital". The *Outline of the 14th Five-Year Plan for National Economic and Social Development* proposes to improve legal norms for identifying monopolies to crack down on monopolies and unfair competition. In February 2021, the Antitrust Commission of the State Council formulated and released the *Antitrust Guidelines for Platform Economy*. The *Guidelines* further clarify the antitrust law-enforcement principles on the basis of the *Anti-monopoly Law of the People's Republic of China* and in combination with the features of the development of a platform economy.[31]

In line with the *Anti-monopoly Law* and the *Guidelines*, the State Administration for Market Regulation investigates and punishes Alibaba Group's monopolistic business. This signals a specific measure taken by regulatory authorities to strengthen antitrust governance and prevent the disorderly expansion of capital, as well as an effective norm to curb the violation of laws and regulations of platform enterprises.

---

[28] www.cac.gov.cn: "To Be a Good Young Internet User in the New Era: Fuzhou 'e-Protection' Youth Online Literacy Micro Course Was Officially Launched". December 9, 2020, http://www.cac.gov.cn/2020-12/09/c_1609081318810794.htm.

[29] www.cac.gov.cn: "The First Xiamen Youth Internet Civilization Contest Was Held". December 15, 2020, http://www.cac.gov.cn/2020-12/15/c_1,609,599,872,016,483.htm.

[30] www.cac.gov.cn: "How to Start the First Lesson? A Model of the 'Rising Star' Project of Internet Literacy in Sanmen County, Taizhou City, Zhejiang Province". March 5, 2021, http://www.cac.gov.cn/2021-03/05/c_1616522429186042.htm.

[31] www.samr.gov.cn: "The Antitrust Commission of the State Council released the *Antitrust Guidelines for Platform Economy*". February 7, 2021, http://www.gov.cn/xinwen/2021-02/07/content_5585758.htm.

5.4 Innovation on Big Screen and Small Screen Integration: New Development Trend of Internet Media

In order to better fulfill their social responsibilities, Internet media continuously make innovations in production and operation, achieve breakthroughs in media convergence, business models, institutional production and technological empowerment, and explore new development paths.

5.4.1 Big Screen and Small Screen Integration Alters Media Ecology

Media convergence means an inevitable trend of media development. With the development of information technology and the change in social environment, grass-roots information content service presents a trend of ecology-based development. Big screen & small screen integration and nestification become a new direction for traditional media convergence.

1. County-Level Media Convergence Center Bolsters the Ecology-Based Services of Grass-roots Information Contents

The construction of county-level media convergence centers effectively integrates grass-roots information resources and promotes ecology-based service of grass-roots information content. As relevant data revealed, by June 2021, the scale of rural Internet users reached 297 million, accounting for 29.4% of all Internet users.[32] Simultaneously, as rural information infrastructure matures, 31 provinces (autonomous regions and municipalities directly under the Central Government) have built government service platforms at provincial, municipal and county levels. Among them, 21 regions have realized five-level service coverage of province, city, county, town and village and initially formed a five-level online service system that features "urban & rural coverage, top-down linkage and clear level". County-level media convergence centers play a key role in ensuring the normal operation of a five-level online service system. The center has functions of both media reporting and government services. It takes public-opinion guidance as the core and serves the masses as the purpose, facilitates "the last mile" of the propaganda & organization front, builds an integrated grass-roots government service platform, and effectively raises grass-roots governance capacity and level. During the COVID-19 pandemic, county-level media convergence departments in various regions give full play to the advantages of media convergence, strictly watch information outlets, update the pandemic data in real time, convey important instructions and various work arrangements of Party organizations and people's governments at all levels on the pandemic prevention & control, and promptly refute rumors and false information. Besides, they

[32] Data Source: *The 48th Statistical Report on China's Internet Development* released by the China Internet Network Information Center.

use multi-media, produce new down-to-earth media products, publicize & popularize knowledge on personal protection, and make outstanding contributions to giving the public a lift and shoring up the pandemic prevention & control.

2. Big Screen & Small Screen Integration Empowers Media Convergence

The interaction between big-screen media represented by traditional TV media and small-screen media represented by the new media platform on mobile phone forms a major trend of media convergence. One method of media convergence is to synchronize the content produced by big-screen media or the content via new media with the corresponding small-screen media, or to transform the brands and programs on big-screen media into new media on small-screen media. Noticeably, under such a model, large-screen media and small-screen media are highly independent. The program "Anchors Talk on CCTV News" launched by CCTV News typifies big screen & small screen integration. In the past year, big screen media and small screen media showed a trend of mutual integration and nestification. The most typical cases are various live shows, such as Hunan SatelliteTV's *Let's go! Brothers* Poverty Alleviation and Agricultural Assistance Live Show and Beijing Satellite TV's "Ordinary Deeds Express the Power of Love" Anti-Pandemic Themed Party. These activities not only retain the features of traditional parties, but also take livestreaming rooms to the parties. The performance and live show are staged simultaneously, and the large screen and small screen are switched freely, which effectively integrates the mechanisms of large screen and small screen.

5.4.2 Media Create New Cross-Field Business Model

Taking advantage of livestreaming to help farmers, mainstream media represented by the CCTV News Channel embark on an innovative path of cross-field cooperation, and the new model of "mainstream media + live commerce" becomes popular. In April 2020, mainstream media engaged in live commerce. New media of www.cctv.com firstly integrated "live commerce" with news broadcasting. Linked via video, CCTV anchors and live-streamers recommended products from Hubei on Internet platforms like CCTV news client-end, Weibo and Taobao. In over-two-hour livestreaming, the number of view on these platforms exceeded 122 million, and the number of like in the livestreaming room exceeded 160 million.[33] The livestreaming finally achieved sales of more than 40 million *yuan* of Hubei agricultural & sideline products and famous and high-quality specialties. This increased sales of products from Hubei and created a new model of mainstream-media-content production. In May 2021, on the day of the launch of "National Consumption Promotion Month", the China Media Group CCTV News Channel and Shanghai Station jointly launched

[33] www.cctv.com: "Recommendation Form of *Thank you for Participating in Team Buying for Hubei*", May 6, 2021, http://news.cctv.com/2021/05/06/ARTImOBN61PMnZDfI73g2Y9V210506.shtml.

"Better Life · Pinduoduo" Ten Billion Subsidy Live Commerce with Pinduoduo. The hosts sold goods on site, with 15 million viewers on the Internet.[34] The New Media Center of *People's Daily* launched the "Revitalization of Old Revolutionary Areas · Support from People" Livestreaming to help the development of old revolutionary areas and do practical work for people in old revolutionary areas, with a total of transactions of more than 730,000. Xinhua News Agency attracted a total of 29.33 million viewers in the first four-hour e-commerce livestreaming under the theme of "beautiful blueprint for thriving business", with the cumulative orders reaching 208,000.

5.4.3 New Media Contents and Technologies Promote Industrial Innovation

Over the past year, the high-tech-backed online cultural industry kept booming, and "cloud exhibition hall", "cloud tourism", and "cloud performance" became popular. In the spring festival of 2021, the National Museum of China launched a series of online activities under the theme of "2021, Becoming Better with the National Museum of China", in which 56 virtual exhibition halls, 100 exhibition webpages and more than 50 short videos were integrated on the cloud. The National Museum of China sincerely invited visitors at home and abroad to enjoy high-quality exhibitions, lectures and cultural relics, and initiated interactive activities such as AI explaining famous cultural relics in the National Museum of China, making special e-cards of cultural relics and producing new-year H5-exclusive signature card. In addition to the digital exhibition hall of the National Museum of China, Digital Palace Museum Applet and Mogao Grottoes Digital Exhibition Center launch similar online exhibitions, which enable Internet users to enjoy a cultural gala without leaving home.

The "5G + UHD" video industry develops rapidly, and technological use case diversifies. The technological integration of 5G with AI, virtual reality and cloud computing spawns new scenes and experiences. In 2021, the CCTV Spring Festival Gala adopted 8 K UHD video live broadcast for the first time. In combination with "AI + VR" naked-eye 3D studio technology, the gala made full use of augmented reality, extended reality and movie special effects technology, via which the audience obtained an immersive experience of changing scenes.[35] In April 2021, at the 4th Digital China Achievements Exhibition, the "5G + 4 K/8 K UHD" video industry in which three major telecom operators (i.e., China Mobile, China Unicom and China

---

[34] *Inner Mongolia Morning News*: "2021 'National Consumption Promotion Month' Was Officially Launched, and CCTV Cooperated with Pinduoduo to Promote Consumption with Live Commerce". May 1, 2021, https://www.thepaper.cn/newsDetail_forward_12518006.

[35] Sina VR: "The Highlight of News of 2021 CCTV Spring Festival Gala: The Use of AI + VR Naked Eye 3D Studio Technology". February 8, 2021, http://vr.sina.com.cn/news/yx/2021-02-08/doc-ikftpnny5795120.shtml.

Telecom) and the National Radio and Television Administration participated became a key display sector.

For mainstream media, new technologies serve as a driving force for the innovation of content and communication models and a major approach for media convergence, transformation and upgrading. On the eve of "Two Sessions" in 2021, "AI Editorial Department" 2.0 of *People's Daily* was launched. The intelligent operating platform integrates AI and 5G technologies and has functions like video editing, poster production and intelligent review, which helps to improve the quality and efficiency of all-media coverage of "Two Sessions". In the coverage of "Two Sessions" in 2021, the intelligent editorial department of Xinhua News Agency launched the first 5G immersive cross-region multi-screen interview in the world, a new cube intelligent studio and an upgraded AI synthesis anchor.[36] New technologies play a part in content production, filtering, distribution and dissemination, and strengthen and enhance the competitiveness of mainstream media in the new media era.

[36] www.ynet.com: "The First 5G Immersive Cross-Screen Interview Actualizes Space Crossing". March 6, 2021, https://t.ynet.cn/baijia/30,471,783.html.

Chapter 6
Cybersecurity Construction

6.1 Outline

Presently, the COVID-19 pandemic rages globally, which accelerates the changes unseen in a century and complicates the international environment. Hegemony in cyberspace poses a threat to global peace and development. Global industrial chain and supply chain are buffeted and threats to traditional security and cybersecurity are intertwined, forming new risks and challenges.

Over the past year, General Secretary Xi Jinping's important instructions of "Four Upholds" on cybersecurity work were implemented in an all-round way. China upheld a holistic approach to national security, implemented new philosophy of development, forged a new landscape of cybersecurity, constantly strengthened the cybersecurity guarantee system and capacity building, reinforced the protection of critical information infrastructure security, enhanced the protection of important data and personal information, and sternly cracked down on cybercrimes according to law. Besides, China continued to optimize the construction of cybersecurity-related technologies, industries and talents and improve the cybersecurity guarantee system, which significantly raised cybersecurity guarantee capacity and continually solidified the awareness of cybersecurity and the skill of basic protection among the entire society.

6.2 Overall Landscape of Cybersecurity Remains Grim

In 2020, a substantial number of new-type Internet products and services emerged, no matter in the fields associated with the pandemic prevention & control or in the fields of production and life, e.g., teleworking, education, medical care and intelligent production. Simultaneously, threats to cybersecurity intensified (e.g., security vulnerabilities, data leakage, cyber fraud and ransom virus), and organized and targeted cyberattacks became prominent, which meant more challenges to cybersecurity work.

Chinese Academy of Cyberspace Studies, *China Internet Development Report 2021*,
https://doi.org/10.1007/978-981-19-9311-4_6

Since 2020, cybersecurity in China has been beset by multiple problems. Specifically, the risks of previous major vulnerabilities remain big. Ransom virus raises more technological requirements and becomes more targeted. Malicious programs of networked intelligent devices are basically spread in the P2P model. The large-scale development of new technologies and applications (e.g., cloud computing, big data, IoT, industrial Internet, blockchain and AI) gives rise to obvious cybersecurity problems. Data security and personal information protection arouse wide attention. Key departments and major industries in China (e.g., government, army, energy, aerospace, telecommunications, media, finance and Internet) become prime targets of APT attacks.

6.2.1 *Traditional Threats to Cybersecurity Continue to Rise*

1. The Number of Newly-Added General Software & Hardware Vulnerabilities has reached a Record High in Recent Years

In 2020, the China National Vulnerability Database (CNVD) added 20,704 general software & hardware vulnerabilities (with a year-on-year increase of 27.9%) and 424 vulnerabilities in cybersecurity products (with a year-on-year increase of 110.9%). In the past five years, the number of newly-added vulnerabilities displayed a significant trend of growth, with an average annual growth rate of 17.6%. In the first half of 2021, the CNVD recorded 13,083 general-purpose security vulnerabilities, with a year-on-year increase of 18.2%.[1] The China National Vulnerability Database of Information Security (CNNVD) received reports of more than 140,000 incident-type vulnerabilities, disclosed around 20,000 general-purpose vulnerabilities, and reported 3,368 general-purpose vulnerabilities.[2] The security vulnerabilities of cybersecurity products call for more attention. The National Computer Network Emergency Response Technical Team/Coordination Center of China (CNCERT) revealed more than 21.764 million times of long-range attacks that used security vulnerabilities to scan and detect domestic hosts and execute codes per day, via sampling monitoring. According to the statistics of the source IP addresses of attacks, attacks mainly came from abroad, accounting for more than 75%.[3] Typical vulnerabilities that attackers used to attack websites were Apache Struts 2 remote code execute and WebLogic deserialization. Typical vulnerabilities that attackers used to attack hosts were "Eternal Blue" and OpenSSL "Heartbleed". The typical vulnerability that attackers used to attack mobile terminals was WebView remote code execute. These typical vulnerabilities once caused serious threats. Albeit they have been exposed for a long time, they are favored by attackers today. It is an important and urgent task to repair these vulnerabilities.

---

[1] Data Source: National Computer Network Emergency Response Technical Team/Coordination Center of China (CNCERT).

[2] Data Source: China Information Technology Security Evaluation Center.

[3] Data Source: National Computer Network Emergency Response Technical Team/Coordination Center of China (CNCERT).

2. The Technologies and Strategies of Ransom Viruses are Constantly Upgraded

In 2020, more than 781,000 ransom-virus software were captured, with a year-on-year increase of 6.8% over 2019.[4] In recent years, ransom viruses turned from extensive attack objects to targeted attack objects, mainly large high-value institutions. Meanwhile, the technological means of ransom viruses were continuously upgraded. The vulnerability-intrusion process and subsequent intranet-horizontal-movement process featured automation, integration, modularization and systematization, with a rapid trend of upgraded attack technology. The strategies of ransom viruses also continued to escalate. Ransom gangs stole encrypted files and sent them back. Then they published partial or complete files on websites or data disclosure websites of dark web to threaten victims with ransoms. For example, a Chinese Internet corporation encountered such attacks from Maze, a ransom gang.

3. Malicious Programs of Networked Intelligent Devices that Use P2P to Spread Become Extremely Active

P2P is a traditional means for the spread of malicious programs, featuring high-speed spread, large-scale infection and difficulty in source-tracing. Malicious program families with networked intelligent devices, such as Mozi and Pinkbot, become extremely active after using P2P. Sampling monitoring demonstrates that in China, the number of networked intelligent devices controlled by P2P reaches up to 22.997 million. In 2020, the scale of botnet-controlled networked intelligent devices expanded. Some large botnets manipulated the controlled end by means of P2P and centralized control.

6.2.2 Threats to Cybersecurity Evolve Quickly in the Era of Digital Transformation

1. Cloud Security

As more businesses are cloudified, the number of cybersecurity incidents or threats on Chinese cloud platforms remains high. In the first half of 2021, the number of large-flow DDoS attacks on the cloud platform accounted for 71.2% of the total number of large-flow DDoS attacks on domestic targets, the number of implanted backdoor

[4] Data Source: National Computer Network Emergency Response Technical Team/Coordination Center of China (CNCERT).

websites on the cloud platform accounted for 87.1% of all domestic implanted backdoor websites, and the number of tampered website on the cloud platform accounted for 89.1% of all domestic tampered websites. Additionally, in most cases, attackers used cloud platforms in China to launch cyberattacks. The number of DDoS attacks on the cloud platform (as control-end) accounted for 51.7% of the total number of DDoS attacks launched by domestic control-end, the number of IP addresses controlled by the control-end of trojan & botnet malware on the cloud platform accounted for 65.1% of the total number of domestic IP address, and the number of malware on the cloud platform accounted for 89.5% of the total number of malware on domestic Internet.[5]

2. Mobile Internet Security

In 2020, the number of newly-added malware on mobile Internet reached *c.* 3.03 million, with a year-on-year increase of 8.5%. According to the statistics of malware, the top three were rogue behavior, fee consumption and information theft, accounting for 48.4, 21.1 and 12.7% respectively. In the first half of 2021, the number of newly-added malware on mobile Internet reached around 868,000, with a year-on-year decrease of 47%, and the top three were the same as those in 2020.[6] Counterfeit Apps comprehensively used targeted delivery, multiple jumps, pan domain name solution and other means to avoid detection. Some hackers made fake Apps and generated QR codes or download links via distribution platforms, and used targeted delivery to cajole the target population into downloading and installing. Some hackers applied a variety of technologies, such as multiple jumps of download links, random changes of domain names and pan domain name solutions to avoid detection. For example, when a counterfeit App download link was detected, it immediately generated new links and spread to avoid detection, which increased the difficulty of governance.

3. Industrial Internet Security

In China, the proportion of industrial control equipment and systems directly exposed to the Internet with high-risk vulnerabilities remains high. In the online security inspection of critical information infrastructure in energy, rail transit and other industries, 20% of production-management systems have high-risk security vulnerabilities, as proved. Besides, the industrial control system becomes an important target for hackers to attack and utilize. Foreign hacker organizations launch targeted attacks on Chinese industrial control video surveillance equipment. As CNCERT's monitoring revealed, in the first half of 2021, a large number of domestic industrial control equipment and systems were exposed to the Internet. To be specific, the equipment included programmable logic controller (51.4%), serial port server (28.3%) and intelligent building equipment (16.0%), and the systems involved power, oil &

[5] Data Source: National Computer Network Emergency Response Technical Team/Coordination Center of China (CNCERT).

[6] Data Source: National Computer Network Emergency Response Technical Team/Coordination Center of China (CNCERT).

gas, rail transit and other key industries, which covered corporate production management, corporate operation management, governmental supervision, industrial cloud platform and so on.

4. Blockchain Security

In 2020, the blockchain industry developed rapidly. In particular, the fast development of DeFi, coupled with the emergence of various projects and applications, posed more severe security challenges to the blockchain ecology. According to the statistics, the economic losses that arose from blockchain security incidents in 2020 far exceeded the total losses from blockchain security incidents in 2018 and 2019, with explosive growth. As relevant reports indicate, there are ten security risks in blockchain infrastructure, i.e., two in authority management, two in key security, one in privacy protection, one in smart contract security and four in security operation & maintenance.[7]

6.2.3 Data Security and Personal Information Protection Become More Prominent

1. The Desensitized Display and Illegal Selling of Citizen's Personal Information Deteriorate

The rapid development of network information technology and application facilitates the access, collection and transmission of citizens' personal information. Sensitive personal information like citizen's ID card number, mobile phone number, home address, educational background and working place is exposed on Internet from time to time. As CNCERT's continuous monitoring confirmed, in 2020, there were 107 cases of desensitized display of citizen's personal information on platforms, such as government services and recruitment & admission announcements, involving nearly 100,000 pieces of desensitized personal information. Additionally, 203 cases of illegal selling of personal information occurred. Among them, the illegal selling of personal information of users in banking, securities and insurance-related industries accounted for *c.* 40% of the total number of illegal data transactions. User data of e-commerce and social platforms, as well as address-list data from the educational industry like universities, training institutions and examination institutions, accounted for 20 and 12% of the total number of illegal data transactions respectively.

2. The Risk of Data Leakage of Networked Database Intensifies

In 2020, CNCERT continuously monitored and reported more than 3,000 important data-security incidents, such as security vulnerabilities in the database of networked information systems, intrusion & control, stealing and illegal selling of personal information, which involved e-commerce, Internet enterprise, medical care, off-campus

[7] Data Source: *Evaluation and Analysis Report on Blockchain Security and Capability 2021* published by the China Academy of Information and Communications Technology (CAICT).

training and other industrial institutions. According to analytical reports, information systems that used mainstream databases like MySQL, SQLServer, Redis and PostgreSQL were frequently attacked. Particularly, brute-force password cracking of databases was the most commonly-seen attack, accounting for 48%. Database deletion, database dragging, malicious code implantation and backdoor implantation happened from time to time, with prominent risks of database vulnerabilities.

3. The Risk of Data Leakage on Applet Rises

In recent years, applet emerged quickly. Yet, the applet exposed more noticeable security risks, especially the risk of the leakage of users' personal information. Centering on five dimensions, i.e., program code security, service interaction security, local data security, network transmission security and security vulnerability, CNCERT carried out security testing on WeChat applets developed by 50 banks in China. The results showed that an average of eight security risks existed in each applet. The risks mainly involved: more than 90% of applets without taking protective measures in cases of the exposure of key information and the input of sensitive information in relation to applet source code, more than 80% of applets without providing personal information collection protocols, more than 60% of applets without encrypting personal information in local storage and network transmission, and a few applets with broken-access risks.[8]

6.2.4 Cybersecurity Risks in Major Industries Remain Serious

1. APT Attacks on Major Industries in China Become More Visible

Attacks on Chinese networks by overseas APT organizations are on the rise.

(1) APT attacks that take advantage of social hot topics to deliver phishing emails occur frequently. Overseas APT attack organizations lure users with relevant social hot topics and working documents like "the COVID-19 pandemic" and "the application for funded projects", and deliver phishing emails to the mailbox accounts of important industrial institutions in China. As monitored, hundreds of email accounts of more than 100 institutions in China were continuously controlled by overseas APT organizations. In particular, many important industrial institutions like military corporations and universities had their hosts controlled because they clicked and executed phishing emails.

(2) Some APT organizations use cyberattack tools to hide in important institutions and equipment in China for a long time and control important targets to steal information. These tools feature powerful functions, complex structures and high concealment. In September 2020, a domestic research institution detected a highly

[8] Data Source: National Computer Network Emergency Response Technical Team/Coordination Center of China (CNCERT).

covert network secret-stealing tool used by the Equation Group on its server. Given that the institution's host was once controlled, the Equation Group started to launch long-term hidden attacks on the institution as early as 2013.

(3) The attack on the supply chain becomes a common approach of APT organizations. Under the background of the COVID-19 pandemic, the demand for teleworking increases significantly, and virtual private network (VPN) becomes one of major technological means for these using teleworking to access their corporate networks. Under such circumstances, some APT organizations disguise trojan files as VPN client-end upgrading packages by controlling VPN servers and distribute the files to important institutions that use VPN servers. As monitored, many overseas APT organizations like Ocean Lotus launched attacks on the Chinese Party and government organs, research institutes and other important departments in this way.

2. The Number of Counterfeit Webpages for Online Administrative Approval Increases Significantly

During the COVID-19 pandemic, administrative approval services shifted to online. At the end of 2020, a large number of counterfeit webpages with the title of "unified corporate license information management system" appeared in China. In November and December, as many as more than 53,000 counterfeit webpages were monitored and detected. Criminals beguiled users into submitting real name, bank card number, balance in card, ID number, reserved mobile phone number and other information on counterfeit webpages. Besides, many counterfeit webpages entitled "nucleic acid testing" and "COVID-19 vaccine appointment" were monitored, whose purpose was to illegally obtain personal privacy information such as user's name, address, ID number and mobile phone number.

3. The Risks of IoV Security

IoV faces risks and challenges in terms of cybersecurity, information security and data security, which poses a great threat to national public security, people's life and property security and personal privacy security and arouses wide attention.

(1) IoV faces the risk of being maliciously controlled. Presently, the IoV system gradually develops the functions of remote diagnosis, dispatching, control, updating and automatic driving of vehicles. For example, there are dispatching commands in intelligent transportation, remote vehicle control based on information platforms, and software remote upgrading of vehicle electronic devices. In these systems, malicious attackers obtain control rights and hijack communication links, which causes cybersecurity problems.

(2) IoV data face the risk of sensitive information leakage. With high intelligence, networked vehicles become the collector, transmitter and carrier of traffic information, up against serious risks in data security. IoV data include not only product data, operation data and communication data of vehicles and equipment, but also traffic sensitive information, user privacy information and corporate internal information of traffic management platforms, information service platform and network operation platform, which need to be strictly managed. In some cases, in order to

improve service quality for users, vehicles with auxiliary or automatic driving functions probably continue to collect the owner's information on privacy data, security data, camera images, short-video clips and vehicle tracking. Now, as the automatic driving function is uncontrollable, the leakage of sensitive information may lead to serious consequences.

6.3 The Work of Cybersecurity Protection and Guarantee is Steadily Promoted

Since 2020, the work of cybersecurity protection and guarantee has been solidified, the policy & standard system of cybersecurity strategy has been improved, and the classified protection of cybersecurity and the protection of critical information infrastructure security have been advanced. Data-security management and personal-information protection have been deeply strengthened, and the password security management system has been quickly optimized. The management of industrial Internet security has been enhanced and the governance of cybersecurity vulnerability has been raised.

6.3.1 Cybersecurity Regulations and Standard Systems Are Continuously Improved

The *Outline of the 14th Five-Year Plan for National Economic and Social Development* clearly proposes to ensure the security of national data, strengthen the protection of personal information, comprehensively reinforce the cybersecurity guarantee system and capacity building, and maintain the security of major infrastructure like water conservancy, power, water supply, oil & gas, transportation, communications, network and finance. Various cybersecurity laws and regulations have been promulgated, such as the *Data Security Law of the People's Republic of China, Personal Information Protection Law of the People's Republic of China, Regulations on the Security Protection of Critical Information Infrastructure, Several Provisions on the Security Management of Automobile Data (For Trial Implementation), Measures for Cybersecurity Review (Draft Revision for Comments)* and *Interim Provisions on the Protection and Management of Personal Information on Mobile Internet Applications (Draft for Comments)*. This signals that cybersecurity laws, regulations and systems in China mature.

Over the years, China has actively promoted the standardization of cybersecurity, constantly strengthened the top-level design of the standardization of cybersecurity, built a complete national standard system for cybersecurity, and attained phased achievements in accelerating the development of key and urgently-needed standards, improving the actual effect of a standard application, and substantially participating in international standardization. In 2020, the National Information Security Standardization Technical Committee released 53 national standards for cybersecurity in a timely manner. By June 2021, a total of 323 national standards for cybersecurity had been issued, which played a fundamental, normative and leading role in supporting the urgent need of national cybersecurity work, safeguarding the vital interests of the majority of Internet users, sustaining the prevention & control of the COVID-19 pandemic, and helping enterprises resume work and production. The compilation and publication of the *White Paper on the Standardization of Cybersecurity Situational Awareness Technology* and the *White Paper on the Standardization of 5G Cybersecurity* provide guidance for the planning and implementation of the standardization in relevant fields. Focusing on hot topics in society, guidelines for cybersecurity standards and practices are compiled and issued, such as *Teleworking Security Protection*, *Guidance for the Prevention of AI Ethics Security Risk*, and *Guidance for Common Problems of Personal Information Protection* as well as the *Handling on Mobile Internet Apps*, which robustly back the standardization in strengthening teleworking security protection and ensuring personal information security during the COVID-19 pandemic.

6.3.2 The Classified Protection of Cybersecurity and the Protection of Critical Information Infrastructure Security Are Continually Advanced

The security of critical information infrastructure means a strategic priority in national security and a top priority of cybersecurity work in China. In recent years, the Office of the Central Cyberspace Affairs Commission has coordinated and promoted the security protection of critical information infrastructure, collaborated with the Ministry of Industry and Information Technology, the Ministry of Public Security and other relevant departments on accelerating the construction of the security guarantee system of critical information infrastructure, strengthened the prime responsibilities of critical information infrastructure operators, and comprehensively solidified cybersecurity work, and effectively ensured the security of nationally critical information infrastructure. The *Critical Information Infrastructure Security Protection Regulations* came into effect on September 1, 2021, which optimized the security-protection measures of critical information infrastructure and clarified the responsibilities of various parties. In October 2020, the Ministry of Industry and Information Technology released the *Notice on Boosting the Work of Cybersecurity Inspection in Telecommunications and Internet Industries in 2020*, which inspected 5G network infrastructure, IP bearer network, management supporting system, public cloud service platform, domain name service system, mobile application store, industrial Internet platform, IoT platform, IoV application service platform and online car-hailing information service platform, reinforced the investigation of hidden cybersecurity risks, reported the inspection results and intensified the rectification. In

November 2020, *GB/T 22,240-2020 Information Security Technology—Classification Guide for Classified Protection of Cybersecurity* formally took effect, which marked the arrival of the 2.0 Era of classified protection. Simultaneously, in important industries like finance and communications, relevant industrial standards on cybersecurity classified protection 2.0 were successively published, which earnestly implemented the cybersecurity classified protection system. In November 2020, the People's Bank of China officially approved the release of *JR/T 0071-2020 Implementation Guidelines for Classified Protection of Cybersecurity of the Financial Industry*, a financial, industrial standard. In December 2020, National Radio and Television Administration approved and issued *GY/T 337-2020 Classification Guide for Classified Protection of Broadcasting Cybersecurity*, an industrial standard of radio & television and network audio-visual. In December 2020, the Ministry of Industry and Information Technology approved and issued *YD/T 3799-2020 Implementation Guidelines for the Rating and Recordation of Telecommunications Network and Internet Network Security Protection.*

6.3.3 The Management of Data Security and the Protection of Personal Information Are Further Strengthened

On January 1, 2021, *The Civil Code of the People's Republic of China* officially came into force, which improved the legal definition of personal information, clarified the connotation, principle and condition of personal information processing, strictly restricted the exemption from liability for processing personal information, guaranteed the rights of correction and deletion of information subjects, enhanced the information security obligations of processors, clearly stated the confidentiality obligations of organs, institutions and personnel, and built a solid firewall for personal information protection. In June 2021, the 29th Meeting of the Standing Committee of the 13th National People's Congress deliberated and adopted the *Data Security Law of the People's Republic of China*, which would be formally implemented on September 1, 2021. As the supreme law in the field of data security, the *Data Security Law of the People's Republic of China* stipulates the obligations and corresponding legal responsibilities of data security protection from the perspectives of data security and development, data security system, data security protection obligations and government data security and openness, and further reinforces data security protection. In August 2021, the 30th Meeting of the Standing Committee of the 13th National People's Congress voted to adopt the *Personal Information Protection Law of the People's Republic of China*, which would be officially executed on November 1, 2021. It clearly stipulates that for information push and business marketing to individuals via automated decision-making, options not specific to personal characteristics or convenient refusal methods should be provided, that processing sensitive personal information such as biometrics, medical care, financial accounts and whereabouts must obtain individual consent, and that applications that illegally process personal information shall be ordered to suspend or terminate the provision of services.

In March 2021, the Cyberspace Administration of China, the Ministry of Industry and Information Technology, the Ministry of Public Security and the State Administration for Market Regulation jointly issued the *Provisions on the Scope of Necessary Personal Information for Common Types of Mobile Internet Applications*, which would come into effect on May 1, 2021. It clarifies the scope of necessary personal information for 39 common types of Apps, including map navigation, online car-hailing and instant message. As it stipulates, mobile Internet app operators shall not forbid user's to use basic function services of the app on the excuse that users do not agree to collect unnecessary personal information. On July 10, 2021, the Cyberspace Administration of China publicly solicited opinions among the public on the *Measures for Cybersecurity Review (Draft Revision for Comments)*, which clearly required that operators who had obtained the personal information of more than one million users must be reviewed in cybersecurity by the Cybersecurity Review Office before they would be listed on the overseas stock market. In August 2021, the Cyberspace Administration of China, the National Development and Reform Commission, the Ministry of Industry and Information Technology, the Ministry of Public Security and the Ministry of Transport jointly released *Several Provisions on the Management of Automobile Data Security (For Trial Implementation)*, which would come into force on October 1, 2021. The *Provisions* focuses on the security risks of personal information and important data in the automobile field, calling on automobile data processors to adhere to the principles of "in-vehicle processing", "no collection of data when driving by default", "accuracy and its applicability" and "desensitized processing" while processing automobile data, so as to reduce the disorderly collection and illegal abuse of automobile data.

6.3.4 Code Security Management System is Quickly Improved

In August 2020, the State Code Administration publicly solicited relevant opinions and suggestions from the public on the *Regulations on the Administration of Commercial Codes (Draft Revision for Comments)*, and comprehensively revised the *Regulations on the Administration of Commercial Codes* issued and implemented in 1999. In September 2020, the Chinese Association for Cryptologic Research Joint Committee for Security Evaluation of Commercial Code Application compiled and released the *Guidelines for Code Application and Security Evaluation of Government Information System (2020 Edition)*, which gave guidelines for the implementation process of code application and security evaluation of government information system in the planning, construction and operation stages, and instructed the construction departments and customer departments of non-secret-related national government information systems to standardize the application and security evaluation of commercial codes. In October 2020, the State Code Administration organized random checks on the use of codes by e-certification service institutions and the business operation of e-government e-certification service institutions, and investigated & punished the violations of laws and regulations, which further standardized the use of codes and the e-certification of e-government in e-certification services and the management of e-signature and data e- message used in governmental activities, and comprehensively improved the guarantee of commercial codes services. In December, the Chinese Association for Cryptologic Research Joint Committee for Security Evaluation of Commercial Code Applications compiled and released five guiding standards (i.e., *Requirements for the Evaluation of Code Application in Information System, Guidelines for the Evaluation Process of Code Application in Information System, Guidelines for the Determination of High Risks of Code Application in Information System, Quantitative-Evaluation Rules for the Security Evaluation of Commercial Code Application*, and *Template for the Security Evaluation Report of Commercial Code Application 2020,* which provided references for relevant departments to carry out commercial code application and security evaluation.

In 2021, new breakthroughs were made in the international standardization of commercial codes in China. SM9 identification encryption algorithm independently developed by China was officially released as an ISO/IEC international standard, and the SM4 block cipher algorithm was officially released as an ISO/IEC standard. In March 2021, the State Administration for Market Regulation and Standardization Administration issued the *National Standard Announcement of the People's Republic of China* (No. 3, 2021). Besides, *Basic Requirements for the Application of Information Security Technology and Information System Codes*, a key standard for national code application and security evaluation, would be formally implemented on October 1, 2021. The standard specified the requirements for code application technology and management in terms of the physical environment, network & communications, equipment & computing, application & data, management system, personnel management, construction & operation and emergency disposal, and possessed great significance for standardizing and guiding the compliance as well as correct and effective application of commercial codes in information systems. In June 2021, the State Code Administration publicly released the updated *Directory of Pilot Institutions for Security Evaluation of Commercial Code Applications* to the public.

6.3.5 The Management of Industrial Internet Security is Constantly Consolidated

The cybersecurity management system of industrial enterprise matures. The R & D of security protection systems for critical equipment, public service platforms and typical scenes of major industries in the industrial Internet field are basically completed, which can be preliminarily promoted and implemented among various industries. In December 2020, the Ministry of Industry and Information Technology issued the *Industrial Internet Innovation and Development Action Plan (2021–2023)*, which aimed to actualize security-enhancement action, boost industrial-Internet-comprehensive-security guarantee-capacity-improvement project, and optimize the classified and graded management system of cybersecurity. In December 2020, the Ministry of Industry and Information Technology issued the *Notice on Carrying out the Pilot Work of Classified and Graded Management of Cybersecurity of Industrial Internet Enterprises*. As initially planned, the pilot work would be carried out in 15 provinces (autonomous regions and municipalities directly under the Central Government), i.e., Tianjin, Jilin, Shanghai, Jiangsu, Zhejiang, Anhui, Fujian, Shandong, Henan, Hunan, Guangdong, Guangxi, Chongqing, Sichuan and Xinjiang, to further improve the classification & grading rules, standards and procedures of cybersecurity of industrial Internet enterprises, as well as the scientificalness, effectiveness and operability of industrial Internet security-series protection specifications.

6.3.6 The Management Level of Cybersecurity Vulnerability is Continuously Raised

In 2020, the CNVD platform was put into trial operation. Open to the public, the CNVD platform enables manufacturers and users to independently obtain and update the function of vulnerability information. In a year, 36 times of emergency responses to major emergency vulnerability incidents were carried out, involving an office automation (OA) system, content management system (CMS) and firewall system. Around 31,000 times of vulnerability incidents were confirmed and addressed. 26 announcements of major security vulnerabilities were timely made to the public,[9] which had an extensive impact and urgently needed to be repaired by end users. These actions effectively hedged against the security risks that probably arose from major security vulnerabilities, evoked the enthusiasm of domestic software & hardware (service) manufacturers to repair their product vulnerabilities, and reduced the security threat of general-use software & hardware product &service vulnerabilities to domestic users.

In June 2020, CNVD officially launched a blockchain vulnerability sub-database, calling on and guiding blockchain security manufacturers, white hat hackers, blockchain enterprises and other parties to jointly participate in the construction of blockchain security ecology, enhance China's capacity to detect, analyze and warn blockchain vulnerabilities and security incidents, improve overall research level and emergency response capability, and provide important technical support and data support for security guarantee of China's blockchain industry. In the past year, the China Information Technology Security Evaluation Center completed the qualification evaluation of information security personnel for more than 10,000 people and the qualification evaluation of information security service for around 100 enterprises.[10]

[9] Data Source: National Computer Network Emergency Response Technical Team/Coordination Center of China (CNCERT).

[10] Data Source: China Information Technology Security Evaluation Center.

In May 2021, the China Information Technology Security Evaluation Center, in cooperation with colleagues in the industry, established the "Jinzhao Cybersecurity Mass Testing Committee" and officially launched the "Jinzhao Security Mass Testing Platform", in a bid to further improve the security protection capabilities of domestic software & hardware products and critical information infrastructure. In July 2021, the Ministry of Industry and Information Technology, the Cyberspace Administration of China and the Ministry of Public Security jointly issued the *Provisions on the Management of Internet Product Vulnerabilities*, which intended to regulate the detection, reporting, repair and release of Internet product security vulnerabilities and prevent cybersecurity risks.

6.4 The Rectification and Governance of Cybersecurity Incidents Are Deepened

Since 2020, the rectification and governance of cybersecurity incidents have been deepened. Relevant departments have continuously actualized the governance of cybersecurity threats to the public Internet, solidly promoted the rectification of prominent cybersecurity problems, and punished illegal behaviors in cyberspace according to law.

6.4.1 The Governance of Cybersecurity Threats to Public Internet is Continuously Carried Out

In 2020, CNCERT coordinated and handled *c.* 103,000 cybersecurity incidents, with a year-on-year decrease of 4.2%. China continued to crack down on computer malware regularly, and successfully closed 386 large-scale botnets in 2020. In the past five years, the number of hosts infected with computer malware in China continually declined and remained at a low infection level, with an average annual decrease rate of 25.1%. In order to root out malware on mobile Internet and effectively block up the main source, CNCERT focuses on coordinating the removal of malicious Apps from domestic registered App disseminative channels. To efficaciously prevent the risks from counterfeit webpages, CNCERT vigorously handles counterfeit webpages in finance and telecommunications. In 2020, CNCERT coordinated with domestic and foreign domain name registration institutions to close more than 17,000 counterfeit webpages. For other counterfeit webpages, the China Cybersecurity Threats Governance Alliance (CCTGA), in conjunction with ten browser manufacturers in China, conducts prompt interception when users visit phishing websites through a collaborative defense pilot. In 2020, the number of prompt interceptions reached 390 million. As sample monitoring suggested, the number of websites implanted with backdoor and tampered websites in China decreased to varying degrees. To be specific, the number of websites implanted in backdoor dropped by 37.3% year on year, the number of domestic government portal implanted with the backdoor dropped by 64.3% year on year, and the number of tampered websites dropped by 45.9% year on year. CNCERT continues to rectify the network resources used for DDoS attacks. The stability of available attack resources in China has been reduced, and the number of domestic attack resources used for cyberattacks has been kept at a low level, which effectively diminishes the attack traffic from China and curbs DDoS attacks from the source. According to the statistics, in 2020, the number of DDoS attacks in China decreased by 16.16%, the total attack traffic decreased by 19.67%,[11] and the proportion of the number of the botnet-controlled terminal in the world steadily decreased to 2.05%.[12]

6.4.2 Prominent Cybersecurity Problems Are Progressively Rectified

Positive results are produced in the governance of illegal collection and use of personal information on Apps. By 2020, the number of downloadable or available active apps in domestic mainstream app stores had totaled 2.67 million, and the number of Android apps and Apple apps had reached 1.05 million and 1.62 million respectively. In order to implement the *Cybersecurity Law of the People' Republic of China*, further standardize the collection of personal information on Apps, and ensure the security of personal information, the Cyberspace Administration of China, together with the Ministry of Industry and Information Technology, the Ministry of Public Security and the State Administration for Market Regulation, continuously cracks down on the illegal collection and use of personal information on Apps, and publicly exposes or removes the Apps with problems against laws and regulations, such as an unauthorized or excessive collection of data as well as compulsory authorization and excessive claims for authorization.

In order to effectively protect citizen's personal privacy and security, from May to August 2021, the Office of the Central Cyberspace Affairs Commission, the Ministry of Industry and Information Technology, the Ministry of Public Security and the State Administration for Market Regulation organized nationwide centralized rectification of cyberspace black industries involving camera peeping, with the focus on the infringement of citizen's personal privacy that citizens detested, such as the illegal use of cameras to peep at personal privacy pictures, trade private videos, and teach peeping technology. Under the guidance of the Office of the Central Cyberspace Affairs Commission, local departments on cyberspace affairs urged various platforms to clean up more than 30,000 pieces of relevant illegal and harmful information, punish more than 5,600 platform accounts and 138 groups, remove more

[11] Data Source: *2020 DDoS Attack Situation Report* jointly released by China Telecom DamDDoS and NSFOCUS Technologies.

[12] Data Source: *DDoS Attacks in Q4 2020* released by Kaspersky Lab.

than 3,000 illegal products, close two websites, investigate networked cameras of more than 700 enterprises, and detect and handle more than 90 hidden dangers related to cameras. The Ministry of Industry and Information Technology comprehensively investigated the security risks of networked cameras and detected 48,000 vulnerabilities of camera equipment (including 46,000 weak-password vulnerabilities), completed the evidence-collection of more than 11,000 vulnerabilities and addressed 1,079 vulnerabilities. The Ministry of Public Security organized public security organs across the country to crack down on illegal and criminal activities according to law, such as providing camera-cracking software tools, launching attack-control on camera equipment and manufacturing & selling eavesdropping and photo-stealing equipment. 17 cases were cracked, 74 suspects were arrested, the right to use more than 26,000 illegally-controlled network cameras was banned, and more than 1,500 sets of eavesdropping and photo-stealing equipment were confiscated. The State Administration for Market Regulation held an administrative guidance meeting for Internet platform enterprises, requiring platform enterprises to strictly perform their main responsibilities, strengthen the governance of counterfeit cameras and other commodities on platforms, and complete comprehensive rectification within a month. With centralized rectification, the vulnerabilities of relevant video-monitoring Apps and camera equipment controlled by black industries were repaired, and the tools in black industries like camera peeping circulated on the Internet, were basically paralyzed. As monitored and proved, 452,000 cameras controlled by black industries for attacks were not freely peeped any longer. Evidently, the centralized rectification achieved positive results.

6.4.3 Illegal and Disorderly Behaviors in Cyberspace Are Punished According to Law

In recent years, with the rapid development of Internet technology, traditional crimes accelerated the shift to Internet-based non-contact crimes. Internet-related crimes such as telecom network fraud, online gambling and infringement of citizen's personal information frequently occurred, which bred other new-type crimes, with the number of Internet-related crimes and the scale of victims continuously growing. National public security organs struck hard and maintained high pressure on Internet-related crimes, and pooled efforts to purify the cyberspace ecological environment. In 2020, national public security organs investigated 6,524 cases that infringed citizen's personal information, with 13,000 suspects arrested, as well as 1,782 cases of hacker attacks and new technology-based crimes, with 2,975 suspects arrested. Besides, national public security organs investigated more than 10,000 cases involving black industries on Internet, arrested 15,000 suspects, seized more than 5.48 million "unregistered mobile phone cards", detected more than 220 million Internet accounts, and timely prevented more than 18.5 million IoT cards from being circulated in the black market. National public security organs initiated the "Internet clean-up campaign

2021" to severely crack down on crimes against citizen's personal information, online gambling, telecommunications network fraud and other illegal activities that harmed citizen's immediate interests, as well as black or gray industries like Internet black cards and black numbers. Since 2021, national public security organs have solved 16,000 cybercrimes, netted 42,000 suspects, thoroughly investigated gangs engaged in black industries that illegally traded black cards and black numbers and operated platforms for manufacturing and receiving verification codes, destroyed hundreds of companies and platforms that provided an illegal diversion, payment and settlement for gambling, fraud and other crimes, and shut down more than ten hacker gangs that produced and developed new-type criminal technologies such as "command trap" and "dark trap". These actions achieved initial results.[13]

In July 2021, in order to prevent national data security risks, safeguard national security and protect public interests, in line with the *National Security Law of the People's Republic of China* and the *Cybersecurity Law of the People's Republic of China*, The Cybersecurity Review Office adhered to the *Measures for Cybersecurity Review* and carried out cybersecurity review on "Didi Chuxing, "Yunmanman", "Huochebang" and "BOSS Zhipin".

6.5 Cybersecurity Industry and Technology Develop Vigorously

In China, as cybersecurity policies and regulations are continually improved and optimized, the standardization of the cybersecurity market gradually progresses, the investment of governmental and corporate customers in cybersecurity products and services steadily increases, and the scale of the domestic cybersecurity market continuously expands.

6.5.1 The Development Trend of Cybersecurity Industry

1. The Cybersecurity Market Comes to a Stage of Rapid Growth

As the International Data Corporation (IDC) predicted, in China, the total expense in the cybersecurity market reached 10.22 billion U. S. dollars in 2021, with a CAGR of 16.8% from 2020 to 2024 as estimated, taking the leading growth rate in the global cybersecurity market.[14] By 2024, the cybersecurity market will grow to 17.27 billion U. S. dollars. In 2020, security hardware continued to dominate the total expense on cybersecurity in China and accounted for 47.2, *vis-à-vis* 32.0 and 20.8% on security service and security software respectively. Government, communications and finance

[13] Data Source: Website of the Ministry of Public Security.

[14] Data Source: *Worldwide Security Spending Guide, 2021v1* published by IDC.

get a chunk of the expenses on the cybersecurity market in China, accounting for *c.* 50% of the overall cybersecurity market.

2. The Proportion of Investment in Cybersecurity Remains Lower than the Average Level in the World

Presently, in China, the proportion of cybersecurity investment in IT investment proves lower than the average level in the world. According to IDC's statistical data, cybersecurity investment accounts for 1.87% of IT investment in China, less than 10% of that of U. S. government (20.4%), lower than the average level in the world (3.74%). This cannot meet the requirements of digital development. In contrast, the federal government of the United States always attaches great importance to investment in cybersecurity. The total IT budget in the fiscal year 2021 reached 92.2 billion U. S. dollars, of which 18.8 billion U. S. dollars was invested to cybersecurity, with an increase of 1.4 billion U. S. dollars over the fiscal year 2020, accounting for 20.4% of the IT budget. As the case stands, the actual spending of the U. S. government on cybersecurity proves higher than its published budget. For example, in the fiscal year 2020, the federal government of the United States applied for a cybersecurity budget of 17.4 billion U. S. dollars, yet it actually spent 18.4 billion U. S. dollars.

3. Product Systems and Industrial Chains Gradually Improve

The technological layout of the cybersecurity industry in China takes initial shape. With the rapid development of the cybersecurity industry, the existing cybersecurity products and services commonly extend from the traditional cybersecurity field to various use cases such as cloud, big data, IoT, industrial control, 5G and mobile Internet. On the basis of the use cases, protected objects and security capabilities of security products and services, cybersecurity products and services in China have covered multiple dimensions, such as basic security, basic technology, security system and security service. The cybersecurity product system gradually improves, with greater industrial vitality. As the cybersecurity industry in China develops fast, the industrial chain now basically takes shape, and the supply–demand relationship becomes relatively clear. In the upstream of the cybersecurity industrial chain, China remains weak in the technological foundation of basic hardware & software systems such as chips, operating systems, databases and middleware, yet forges complete technological capability in basic fields such as engines, algorithm and rule bases. In the midstream of the cybersecurity industrial chain, cybersecurity products and services in China develop stably and form a complete technological layout. In the downstream of the cybersecurity industrial chain, institutions in the Party, government and military organizations as well as enterprise users are the main consumers of cybersecurity products and services.[15] Figure 6.1 shows the classification archi-

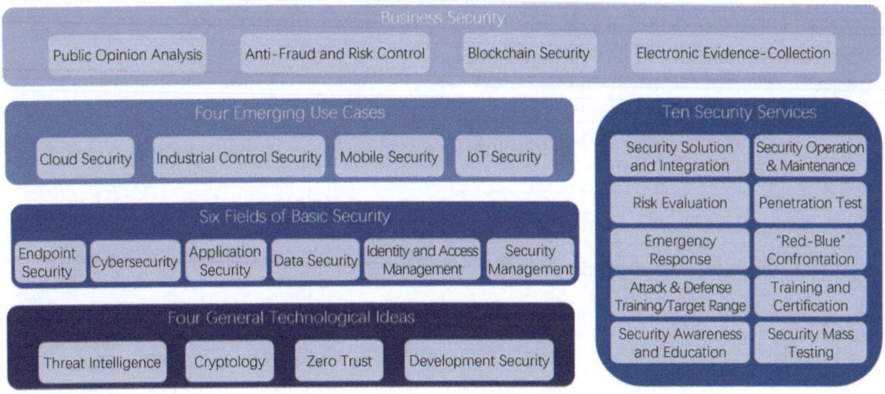

Fig. 6.1 The classification architecture of the cybersecurity market in China

tecture of the cybersecurity market in China, and Table 6.1 shows the classification of cybersecurity products and services in China.

4. Cybersecurity Signals a Direction in Investment

In 2020, there were more than 60 investment & financing transactions in cybersecurity-related enterprises in China, with an investment & financing scale of around 6 billion *yuan*, involving more than 50 cybersecurity enterprises and covering various niche fields like cloud security, identity management, threat detection, industrial control security, data security, IoT security and blockchain security. Some cybersecurity enterprises completed two-round financing in a year. From January to July 2021, there were 67 unlisted investment & financing transactions in the cybersecurity sector in China, with the disclosed amount exceeding 9.5 billion *yuan*. The number of investment & financing events increased by 86%, and the amount of investment & financing increased by nearly six times compared with the same period a year ago. In terms of amount, nearly 40% of investment & financing transactions raised more than 100 million *yuan*, nearly 20% of transactions raised more than 200 million *yuan*, and the number of financing transactions with more than 100 million *yuan* increased by *c*. 120% over the same period in 2020. With respect to the financing stage, the investment institutions paid equal attention to the early and middle & late stage projects in the field of cybersecurity. The middle & late stage projects (including round B and above, listing and fixed increase, and so on) accounted for 47.8% of the total financing events, slightly higher than the early-stage projects (specific financing rounds were not disclosed in some projects). In regard to the niche track, top three niche tracks in the early project financing with public attention were cloud security, security management & operation, industrial control security (tied for the second) and data security, among which cloud security track had the largest number of financing transaction (7). Top three niche tracks in the middle & late stage project financing with

[15] Data Source: *White Paper on China Cybersecurity Industry 2020* released by the China Academy of Information and Communications Technology (CAICT).

Table 6.1 The classification of cybersecurity products and services in China[16]

Cybersecurity products and services	Classification	Sub-classification
The fields of basic security	Network and basic architecture security	Firewall/UTM/2G firewall, Internet Behavior management, VPN/Encryption machine, Intrusion detection & Defense, Network isolation and one-way import, Anti-virus gateway, Cybersecurity review, Anti-DoS attack (Device), Network access and control, Protection against APT, Network traffic analysis, Security review, DNS security, Application delivery/loading balance, Fraud-prevention technology, SD-WAN
	Endpoint security	Protection against malware, Terminal security management, Terminal detection & response, Host/Server reinforcement
	Application security	Web application firewall, Web application security scanning and monitoring, Webpage anti-tampering, Email security, API security
	Data security	Data security governance, Personal privacy protection, Database security, Secure database, Data desensitization, Protection against data leakage, Electronic document management & encryption, Storage backup & recovery
	Identity and access management	Identity authentication & authority management, Operation & maintenance review bastion host, Privileged account management, Digital certificate, Hardware authentication

(continued)

[16] Data Source: *China Cybersecurity Industry Analytical Report* 2020 released by the China Cybersecurity Industry Alliance.

Table 6.1 (continued)

Cybersecurity products and services	Classification	Sub-classification
	Security Management	Security management platform/situation awareness, Log analysis & review, Cybersecurity asset management, Compliance inspection tool, Security baseline and configuration management, Vulnerability evaluation & management, Threat management, Security orchestration automation and response, Security management integrated machine, City-level security operation
General technological ideas	Threat intelligence	
	Cryptology	
	Zero trust	
	Development security	
Emerging use cases	Cloud security	Cloud operating system, Virtualized security products, Container security, Micro isolation, Cloud workload protection platform, Cloud security resource pool, Virtualization & super-integration, Cloud desktop, Cloud identity management, Cloud anti-DDoS, Cloud WAF
	Mobile security	Mobile terminal security, Mobile application security, Mobile security management
	IoT security	IoV security, Video private network security and others
	Industrial control security	
Business security	Business security	Public opinion analysis, Anti-fraud and risk control, Blockchain security, Electronic evidence-collection
Security service	Security service	Security solution and integration, Security operation & maintenance, Risk evaluation, penetration test emergency response, "Red-Blue" confrontation, Attack & defense training/target range, Training and certification, Security awareness and education, Security mass testing

public attention were industrial control security, data security and privacy computing (tied for the second) and identity security and security service (tied for the third), among which industrial control security track had the largest number of financing transaction (10).[17]

6.5.2 *The Construction of Cybersecurity Industrial Park is Continuously Promoted*

The construction of the National Cybersecurity Industrial Park achieves initial results. In June 2020, the National Cybersecurity Industrial Park (Changsha) was officially launched, and the first-batch introduced enterprises and projects were signed. The Park was officially approved in December 2019, with the high-tech zone as the core area. Currently, 184 key cybersecurity enterprises have entered the core area, including 52 above-designated-size enterprises, 123 high-tech enterprises and 6 listed enterprises. The industrial chain basically takes shape, covering basic hardware, basic software, information security service, industrial Internet security and other fields. The National Cybersecurity Industrial Park Haidian Park, Tongzhou Park and Economic Development Zone Information and Innovation Park have been established. More than 100 enterprises have settled in the three parks. Among them, Tongzhou Park was officially opened in November 2020. As national cybersecurity high-end industrial cluster demonstration base and cybersecurity leading talent training base, Tongzhou Park undertakes the tasks of the cluster development of the cybersecurity industry, the innovation and application of products & services and the implementation of cybersecurity talent training programs.

6.5.3 *Industrial Organizations Foster the Development of Cybersecurity Industry*

The Cyber Security Association of China centers on the protection of critical information infrastructure, data security and personal information security and carries out standard formulation, talent training and industrial development, so as to serve the overall interests of China and boost the development of the cybersecurity industry. In June 2020, the Cyber Security Association of China released the *Statistical Report on Cybersecurity Industry in China 2020* for the first time. In August 2020, the Cyber Security Association of China organized a review in the field of cybersecurity in China to recommend the Outstanding Engineer Award of China International Science Exchange Foundation in 2020 (the Fourth Session). At the beginning of 2021, the working groups and research groups were set up in the fields of data governance,

---

[17] Data Source: Data of the National Industrial Information Security Development Research Center, by August 2021.

IPv6 security, financial security, juvenile protection, AI security and cybersecurity talent. In February 2021, the Cyber Security Association of China set up a special working group for personal information protection, which took concerted action with the Cyberspace Administration of China, the Ministry of Industry and Information Technology, the Ministry of Public Security and the State Administration for Market Regulation, and undertook the work related to the acceptance of reports on the collection and use of personal information in violation of laws and regulations on Apps and industrial self-discipline. Since May 2021, the Cyber Security Association of China, working with the Office of Shanghai Cyberspace Affairs Commission, the Office of Hubei Cyberspace Affairs Commission and the China Internet Network Information Center, organized the activity of "Celebrating the Centenary of the Communist Party of China · Cybersecurity China Trip" in Shanghai and Wuhan. On such topics as cybersecurity, data security, satellite Internet, disaster preparedness, talent training and juvenile protection, relevant organizations exchanged opinions on the opportunities and challenges in the cybersecurity industry in the "14th Five-Year Plan".

The Internet Society of China attaches attention to the Internet protection and Internet information governance of the minors, continues to improve relevant measures, and accelerates the formulation and implementation of standards such as the *Framework of the Prevention System of Harmful Information Against the Minors of Internet Operators* and *Guidelines for Labeling Artificial Intelligence Data for Content Security*.

The China Cybersecurity Industry Alliance (CCIA) continues to conduct the selection of excellent cybersecurity solutions and innovative products, selects top 50 enterprises as well as enterprises with growth potential of CCIA cybersecurity industry from the perspective of industrial competitiveness, and organizes member units to participate in the formulation of national standards, which play a positive role in promoting the innovation-driven development of cybersecurity industry and introducing excellent cybersecurity enterprises.

The Zhongguancun Trusted Computing Industry Alliance actively promotes the development of cybersecurity and trusted computing industries in China, by organizing publicity & education activities, building sharing platforms and formulating association standards. In October 2020, the Zhongguancun Trusted Computing Industry Alliance participated in the establishment of Cybersecurity Graded Protection System 2.0 and Trusted Computing 3.0 Research & Demonstration Base, and built support platforms for technological research, adaptation testing, typical demonstration and achievement display. In April 2021, the Alliance hosted the 8th "4.29 Capital Cybersecurity Day" Series of Publicity Activities and "The Innovation and Development Forum on Graded Protection 2.0 and Trusted Computing 3.0".

6.6 Cybersecurity Talent-Training, Publicity and Education Are Continually Promoted

Improving public awareness of cybersecurity and cultivating cybersecurity talents form an important foundation of cybersecurity. In recent years, as the publicity and education of cybersecurity keep deepening, the awareness of cybersecurity in the whole society is significantly improved, and great progress is made in talent construction. Yet, the problem of the shortage of cybersecurity talent in China remains unsolved, posing an urgent task to strengthen cybersecurity education and training.

6.6.1 The Demand for Cybersecurity Talent Expands and the Supply Remains Insufficient

1. The Demand for Cybersecurity Talent in China Enlarges Rapidly, and the Reserve of Practical Talent Remains Insufficient

According to big data of the "Zhaopin" platform, in the first quarter of 2021, the demand of enterprises for cybersecurity talents increased year on year for three consecutive months, and the growth rate enlarged month by month. In terms of organizations that recruit cybersecurity talents, the demand of private enterprises for cybersecurity talents accounted for 57.2% of the total cybersecurity-talent recruitment, and the proportion exceeded 50% in the recent three years.[18] Presently, the demand for talents in the security industry remains robust. However, the training of cybersecurity talents takes a long cycle in years of systematic experience in practical business cases. Now, in addition to the insufficient supply of cybersecurity talents in China, the quality of talents proves unsatisfactory. There is an acute shortage of high-level professionals with keen security awareness and practical experience in cyber attack & defense, who can analyze and solve complex security problems and high-level security threats.

2. The Utilization Rate of Qualified Talents in the Cybersecurity Field Remains Low, and the Curriculum Settings of Related Majors Disaccord with the Actual Needs of Enterprises

In the 15 college majors related to cybersecurity, the average proportion of graduates engaged in computer or Internet technology R & D is 59%. Among them, only 10% of professional talents directly engage in the field of cybersecurity. There is a big gap between the curriculum settings of related cybersecurity majors and the actual needs of enterprises. Besides, overall competitiveness like industrial salary and personal growth space stays at the average level, which does not appeal to talents.

[18] Data Source: *Research Report on Cybersecurity Talent Market in 2021* jointly released by Qi'anxin Industrial Security Research Center and Zhaopin, https://www.qianxin.com/news/detail?news_id=1060.

6.6.2 *The Cultivation of Cybersecurity Talents is Constantly Strengthened*

1. The Demonstration Project of Building First-class Cybersecurity Colleges Continuously Advances

General Secretary Xi Jinping's important instruction and spirit of "making great efforts, increasing investment, inviting top-level teachers, compiling high-quality textbooks, enrolling excellent students and building first-class cybersecurity colleges" are deeply implemented. In 2017, the Office of the Central Cyberspace Affairs Commission and the Ministry of Education jointly organized and implemented the Demonstration Project of Building First-Class Cybersecurity Colleges. In 2017 and 2019, 11 universities were selected for the demonstration project in two batches. Since the implementation of the demonstration project, under the guidance of relevant departments and with the support of local governments, 11 universities have attained remarkable results in terms of policy guarantee, capital investment, basic condition, teaching resources, student cultivation and scientific-research innovation, and made significant progress in building cybersecurity colleges, with prominent demonstration and driving effect. The policy guarantee system continuously improves, the level of teaching & office infrastructure significantly rises, the faculty team continually enlarges, and the scale of teachers rapidly grows, with an increase of more than 40% in the number of selected demonstration projects. The enrollment scale increases significantly, with sustainable innovation on the cultivation model. Currently, there are more than 13,000 full-time students in 11 cybersecurity colleges. The industrial integration of education and technology forms distinctive strengths, and the capacity of scientific research & innovation constantly improves.

Spurred by the demonstration projects, the discipline construction of domestic cybersecurity colleges makes considerable progress. 45 universities have set up independent cybersecurity colleges, with 33 doctoral and 42 master degree programs in cybersecurity, which vigorously promotes the development of the cybersecurity discipline. More than 170 universities in China have set up five cybersecurity-related majors, i.e., cybersecurity, information security, information countermeasure technology, confidentiality technology and cybersecurity and law enforcement, with a total of 271 courses. 27 cybersecurity majors are recognized as national first-class undergraduate major construction sites, with more than 20,000 cybersecurity graduates each year.

2. The Construction of the National Cybersecurity Talent and Innovation Base Accelerates

In September 2020, the cybersecurity colleges of Wuhan University and Huazhong University of Science and Technology, two "first-class cybersecurity college building demonstration project universities", were transferred to the National Cybersecurity Base as a whole, so as to jointly build a world-class cybersecurity college, together with more than 1,300 students and more than 100 teachers and staff. National Cybersecurity Base, Wuhan University and Huazhong University of Science and Technology jointly build an integrated innovation system of "government, production, university, research and application", which supports Wuhan University to build the Cybersecurity Virtual Simulation Experimental Teaching Center, International Cyberspace Governance Research Base, Key Laboratory of the Ministry of Education, Hubei Provincial Cybersecurity Research Center (Hubei Cybersecurity Think Tank) and other platforms, and supports Huazhong University of Science and Technology to build the Innovation Research Institute, Hubei Provincial Engineering Research Center for Big Data Security and other platforms. More than 100 scientific research projects are jointly carried out.

Under the guidance of the Office of the Central Cyberspace Affairs Commission, the "Cybersecurity 10,000 Talents Training Funding Program" is jointly staged by Wuhan Municipal People's Government, the China Internet Development Foundation, the China Information Security Certification Center, the China Information Technology Security Evaluation Center and the National Computer Network Emergency Response Technical Team/Coordination Center. It funds the institutions that launch cybersecurity training for staff of Party and government organs and enterprises and institutions as well as college students in the National Cybersecurity Base, and awards outstanding students. On March 20, 2021, the National Cybersecurity Talent and Innovation Base Training Center was established, and the "Cybersecurity 10,000 Talents Training Funding Program" was officially launched, with 15 training organizations shortlisted. By the end of July 2021, the National Cybersecurity Base Training Center had carried out more than 70 training courses and on-site teaching activities, and the number of trainees had exceeded 5,000.

3. Cybersecurity and Other 15 Fields are Included in the National Security Education, and the Curriculum Settings of Vocational Education Are Optimized

In September 2020, the Ministry of Education issued the *Guiding Outline for National Security Education in Universities, Middle Schools and Primary Schools*, which stated that the main contents of cybersecurity included network infrastructure, network operation, network service and information security, which constituted the basis for ensuring and promoting the healthy development of information society. To safeguard cybersecurity, relevant departments must implement the idea of "national security depends on cybersecurity, and modernization depends on informatization", strengthen law-based cyberspace governance, technological innovation and international cooperation, and forge the awareness of cyberspace sovereignty.[19] In March 2021, the Ministry of Education issued the *Professional Catalogue of Vocational Education (2021)*, setting up nearly ten secondary vocational-related majors (e.g., network information security), more than ten vocational related-majors (e.g., information security technology application), and five professional undergraduate-related

---

[19] Ministry of Education: "Notice on Printing and Distributing the *Guiding Outline for National Security Education in Universities, Middle Schools and Primary Schools*". October 20, 2020, http://www.moe.gov.cn/srcsite/A26/s8001/202010/t20201027_496805.html.

majors (e.g., information security and management) in the field of cybersecurity, and forming more than 3,600 professional courses in majors like information security technology and application in vocational colleges across the country.

4. Cybersecurity Platforms Gather Leading Industrial Talents

Over the past year, various regions actively utilized cybersecurity platforms to gather and train talents in this regard. Guangdong Province implemented provincial-level R & D programs in major fields, reinforced the construction of Peng Cheng Laboratory, introduced a group of top expert teams in the field of network information security led by well-known academicians, deployed and promoted core technological breakthroughs like "National Cyber Range" and "Future Regional Network", and cultivated a batch of high-level cybersecurity talents. In July 2020, Jiangsu Cybersecurity (Wuxi) Training Base was completed and put into use. The base will fully play a demonstrating and driving role and become a "new cradle" for the development of cybersecurity talents as planned. In September 2020, the Tibet Autonomous Region Cyberspace Talent Training Base was organized, which marked a solid step in local cyberspace talent training and discipline construction. In April 2021, Zhejiang Province "2021 West Lake Cybersecurity Conference" set up a forum on the construction of cybersecurity talent teams, in which guests put forward suggestions and implementation plans on the construction of cybersecurity talent teams, so as to bolster the cooperation of the upstream & downstream industrial chains in "government, university, enterprise and bank", and provide new ideas for the construction of cybersecurity talent teams in China.

6.6.3 With Diverse Cybersecurity Publicity and Education Activities, Cybersecurity Awareness is Significantly Enhanced

1. In the Forms of the National Cybersecurity Publicity Week and National Security Education Day for All Citizens, All Citizens Participated in Building the Cybersecurity Defense Wall

Under the theme of "Cybersecurity Serving and Depending on the People", the National Cybersecurity Publicity Week was firstly launched in 2014. It has been staged for seven times so far, in the third week of September every year. In the National Cybersecurity Publicity Week, relevant departments across the country use innovative methods to expand the scope of publicity, highlight the knowledge, interaction and experience in addressing problems like personal information leakage and online fraud that concern the public, popularize the general knowledge of cybersecurity in a simple and interesting way, and improve the protection skills of the public, which evoke strong response in the whole society. On April 29, 2021, Beijing hosted the 8th Capital Cybersecurity Day in response to the annual theme of "Welcoming Beijing Olympic Winter Games". Focusing on the topic of "Undertaking

Cybersecurity Responsibility and Sharing Cyber Life", activities were planned and organized to promote "public education, industrial development, talent cultivation and cultural advancement". A series of major activities in the 2021 National Cybersecurity Publicity Week was held in Xi'an. Cybersecurity publicity and education were carried out in an innovative way. Combining online forms (e.g., image-text, livestreaming, short video, MV, puzzle game and cloud course) with offline channels (e.g., themed exhibitions, tour lectures, skill competitions and community publicity), policies on cybersecurity were deeply interpreted, knowledge on cybersecurity was widely popularized, and model teaching was provided, which solidified the public awareness of cybersecurity.

2. Various Regions Actively Launch the Publicity and Education of Anti-Telecom Network Fraud

In 2020, the National Anti-Fraud Center blocked a total of 1.6 million fraudulent domain name websites, successfully dissuaded 8.7 million people from being cheated, and saved a cumulative loss of 187.6 billion *yuan*. At the 315 Evening Gala in 2021, four issues, i.e., traps on the mobile phone for the elderly, CV disclosure, face-recognition vulnerability and false information in browsers, aroused widespread attention to telecom fraud and cybersecurity among the public. On April 27, 2021, Publicity Department of the CPC Central Committee and the Ministry of Public Security jointly launched the "Mobilizing the Mass for Anti-Fraud" National Anti-Fraud and Prevention Against Fraud Publicity Campaign and simultaneously organized the "Centralized Publicity Month for Cracking down on Telecom Network Fraud Crimes" nationwide. Local governmental departments actively carried out publicity and education to prevent telecom network fraud.

3. The Survey Demonstrates that Internet Users' Satisfaction with Cybersecurity Rises, with Their Awareness of Cybersecurity Remarkably Enhanced

In July 2020, 135 national cybersecurity industry associations and relevant social organizations jointly initiated a survey of Internet users' satisfaction with security. The survey divided the value of Internet users' sense of security into five levels, i.e., "of high security", "of security", "of common security", "of no security" and "of danger". As the survey report showed,[20] 50.04% of Internet users considered cyberspace in China "of high security" and "of security" in 2020. In 2020, the proportion of Internet users who viewed cyberspace in China as "of common security" was 2.85% higher than that in 2019. The proportion of Internet users who regarded cyberspace in China as "of no security" and "of danger" slightly decreased. Relevant governmental departments made achievements in strengthening cybersecurity regulation, with the regulation widely recognized. 54.96% of the practitioners believed that the regulation was "enormously augmented" or "strongly augmented".

[20] Data Source: *National General Report on 2020 Internet Users' Satisfaction with Cybersecurity*, https://mp.weixin.qq.com/s/iVUE2MygFOcS5vM_QgB4KQ.

Chapter 7
Rule-of-Law Construction in Cyberspace

7.1 Outline

Cyberspace means a new area for comprehensively promoting law-based governance. For a long time, China has adhered to the law-based governance of cyberspace, law-based construction of websites and law-based use of the Internet, which continuously further the rule of law construction in cyberspace. In August 2021, the CPC Central Committee and the State Council issued the *Implementation Outline for the Construction of a Government under the Rule of Law (2021–2025)*, which further promoted the working deployment of the rule of law construction in cyberspace, required relevant departments to promptly research legal systems in the digital economy, ITFIN, AI, big data and cloud computing, vigorously improve weak industries, and ensure the healthy development of new business forms and models with the complete legal system and satisfactory governance effect. Over the past year, law played a prominent role in cyberspace governance, the legislative system in cyberspace continuously improved, and the law-enforcement in cyberspace intensively expanded, with positive progress made in law-popularization and remarkable results achieved in cyberspace governance. The top-level legislative design for data security and personal information protection takes shape. The *Data Security Law of the People's Republic of China* has been promulgated and implemented. The *Personal Information Protection Law of the People's Republic of China* has been issued. *The Civil Code of the People's Republic of China* gradually improves the rules on privacy & personal information protection, and systematically strengthens legislative, law-enforcement and judicial work in the fields of the digital economy, platform competition, online transaction, online livestreaming and smart judiciary. Online publicity and education on law-based governance of cyberspace further expand, and Internet user significantly raises their legal literacy and enhance their legal awareness. As the *Regulations on Ecological Governance of Internet Information Content* is deeply implemented, a comprehensive cyberspace governance system gradually improves.

Chinese Academy of Cyberspace Studies, *China Internet Development Report 2021*,
https://doi.org/10.1007/978-981-19-9311-4_7

7.2 The Legislation on Cybersecurity Continues to Advance and the Top-Level Design of Data Protection Takes Initial Shape

As an important field to be restricted in cyberspace governance, security has been a central task in legislation, judicature and law-enforcement. In 2021, the special law on data security was officially introduced, and a series of documents related to security management were released successively, with tireless efforts taken in law-enforcement and judicature. The cybercrime governance campaign was further advanced, and targeted efforts were made to crack down on the use of information networks to commit crimes. Cybersecurity law-enforcement was continually launched. Multiple departments jointly combated illegal and criminal activities that endangered data security, with long-term mechanisms and comprehensive governance systems gradually established.

7.2.1 The Legislation on Basic Data Security is Continuously Promoted

With the convergence of information technology and human production and life, all types of data grow rapidly, which has a far-reaching impact on economic development, social governance and people's life. Data security is a major issue that concerns national security and socio-economic development. On June 10, 2021, the *Data Security Law of the People's Republic of China* (the *Data Security Law* for short) was adopted at the 29th Meeting of the Standing Committee of the 13th National People's Congress and formally implemented on September 1, 2021.

The *Data Security Law* serves as a basic law in the field of data regulation and an important law in the field of national security in China. As a law with diversified legislative purpose, extensive adjustment scope and prominent local features, the *Data Security Law* represents the governance needs of the development and digital transformation of the data element market in China, and marks a new start for future data governance. It forms various basic systems for data security management, such as data grading & classification management, risk evaluation, monitoring & early warning and emergency response, stipulates measures to promote data security and development, establishes the system to ensure government data security and enlarge open government data, clarifies the obligations of organizations and individuals for data security protection in data security-related activities, and implements data security protection responsibilities. Simultaneously, the *Data Security Law* further clarifies the regulation mechanism of data security in China, stipulates the leadership of the central national security leading organs in national data security work, and establishes the Coordination Mechanism of National Data Security Work. The competent departments of industry, telecommunications, transportation, finance, natural resources, health, education and science & technology shall assume the responsibility for data security regulation in their own industries and fields. Public security organs and national security organs shall, in accordance with the provisions of laws and administrative regulations, assume the responsibility of data security regulation within their respective scope of responsibility. In line with the provisions of laws and administrative regulations, national cyberspace affairs departments take charge of coordinating network data security and related regulatory work.

7.2.2 *Cybersecurity Review System is Continually Improved*

In order to safeguard national security and ensure network-data security, the Cyberspace Administration of China, together with the National Development and Reform Commission, the Ministry of Industry and Information Technology, the Ministry of Public Security and other departments, builds the national cybersecurity review mechanism, establishes the Cybersecurity Review Office, launches cybersecurity review, and solicits public opinions on the *Measures for Cybersecurity Review (Draft Revision for Comments)* in July 2021, in accordance with the *National Security Law of the People's Republic of China*, the *Cybersecurity Law of the People's Republic of China* and other relevant laws and regulations, under the leadership Office of the Central Cyberspace Affairs Commission. The *Draft Revision* makes important adjustments in the scope of the declaration, member units of the mechanism, special review time and other aspects. In addition to regulating critical information infrastructure operators, "data processors" is added as the main entity of regulation. Besides, the *Draft Revision* raises new requirements for the review of companies listed overseas. In particular, the newly-added "Article 6" clearly states that "operators who get access to the personal information of more than one million users must be reviewed in cybersecurity by the Cybersecurity Review Office before they will be listed on the overseas stock market". Noteworthily, "Article 24" of the *Data Security Law of the People's Republic of China* stipulates that "China shall establish the data security review system to conduct the national security review of data processing activities that affect or may affect national security".

In July 2021, the Cybersecurity Review Office started a cybersecurity review against "Didi Chuxing" app. As verified, "Didi Chuxing" app has serious problems in collecting and using personal information in violation of laws and regulations. App Store was notified to remove "Didi Chuxing" app, and Didichuxing Technology Co., Ltd. was required to seriously rectify its own problems in accordance with legal requirements and relevant national standards, so as to effectively ensure the security of users' personal information. Simultaneously, the Cybersecurity Review Office issued an announcement on launching the cybersecurity review against "Yunmanman", "Huochebang" and "BOSS Zhipin".

7.2.3 *The Protection of Critical Information Infrastructure is Constantly Enhanced*

In July 2021, the State Council promulgated the *Regulations on the Protection of Critical Information Infrastructure Security*, which would come into force on September 1, 2021. The *Regulations* intends to implement relevant requirements of *the Cybersecurity Law of the People's Republic of China* and provide strong legal guarantees for deepening the protection of critical information infrastructure security in China. In July 2020, the Ministry of Public Security issued the *Guiding Opinions on Implementing the Cybersecurity Multi-Level Protection System and Critical Information Infrastructure Security Protection System.* This signals the implementation of the graded protection system 2.0 and explicitly raises requirements for the protection of critical information infrastructure for the first time, with high applicability. The document proposes to strengthen the protection of important data and personal information, take multiple technological measures to ensure the security of important data throughout the lifecycle, and build a cybersecurity technological protection system that centers on cryptography, trusted computing, AI, big data analysis, etc. The National Information Security Standardization Technical Committee releases the drafts for comments of two standards, the *Method for Determining the Boundary of Critical Information Infrastructure of Information Security Technology* and the *Method for Evaluating the Security Protection Capability of Critical Information Infrastructure of Information Security Technology*, which put forward standards and requirements for determining the boundary and evaluating the security protection capability of critical information infrastructure.

7.2.4 *The Special Rectification Against Cybercrimes is Launched*

In recent years, with the rapid development of information networks, traditional criminal activities crept into cyberspace. Different from traditional offline crimes, information network-based crimes feature a young subject of crime, loose organizational structure, detailed criminal-activity division and soft violence in criminal means. In 2020, telecom network fraud cases occurred frequently, the number of online gambling crimes rose significantly, and the black & grey industrial ecosystem of online crime emerged and developed. According to the work report published by the Supreme People's Court in Two Sessions in 2021, in 2020, the national judiciary concluded 33,000 criminal cases of telecom network frauds, Internet-based pyramid schemes, online gambling, Internet hackers, Internet rumors and Internet violence, heard the cases of Chen's extraordinary cross-border telecom fraud and Wang's buying and selling other's social platform accounts, and severely punished criminal subjects who violated citizen's property and personal information. In 2020, national public security organs launched the "Internet clean-up campaign 2021" to crack down on online black industry crime clusters, severely punished illegal and criminal activities that provided technological support or services for telecom network fraud and online gambling, investigated 4,453 criminal cases, arrested 14,311 illegal suspects, and shut down more than 220 million Internet accounts involved.

7.3 The Legislative Process of Personal Information Protection Accelerates, with Law-Enforcement and Regulation Tightened

With the promulgation and implementation of *The Civil Code of the People's Republic of China*, the legal system of personal information protection in China continues to improve. The *Personal Information Protection Law of the People's Republic of China* has been deliberated and adopted, providing strong support and guidance for the specific implementation of the legal system of personal information protection.

7.3.1 The Civil Code *Ameliorates the Rules on Privacy and Personal Information Protection*

The Civil Code makes special provisions on privacy and personal information protection, clarifies the scope of the legal protection of privacy and personal information, and completes the top-level legislative design of personal information protection. Chapter 6 or "Personality Rights" of *The Civil Code* stipulates "the right to privacy and personal information protection", provides a remedy basis for the right to privacy and personal information protection in the field of civil private law, complements public-law protection, and establishes basic rules and principles for judicial adjudication. On the one hand, *The Civil Code* fully defines privacy, clearly prohibits privacy infringement, and plays a normative and guiding role. On the other hand, *The Civil Code* confirms the legal protection of personal information of a natural person, specifies the principles of legality, legitimacy, necessity and appropriateness in the processing of personal information, and states the conditions for processing personal information, including obtaining the consent of the natural person or the guardian. Additionally, "Personality Rights" of *The Civil Code* further stipulates that civil subjects enjoy the right to informed consent, the right to consult (visit), the right to copy (or obtain copies), the right to correct, and the right to delete. These provisions give citizens the right to personal information protection in a more comprehensive way.

7.3.2 The Special Legislation on Personal Information Protection is Deliberated and Adopted

General Secretary Xi Jinping reiterates on various occasions that we should adhere to the principle that cybersecurity serves the people and depends on the people, safeguard the security of personal information, and ensure citizens' legitimate rights and interests in cyberspace, which put forward clear requirements for strengthening personal information protection. On August 20, 2021, the Standing Committee of the 13th National People's Congress voted to adopt the *Personal Information Protection Law of the People's Republic of China*, which would come into effect on November 1, 2021. Its relevant provisions correspond to these in *The Civil Code*, *The Criminal Law*, *The Cybersecurity Law* and other laws, forming a legal-liability framework in civil, administrative and criminal dimensions. The *Personal Information Protection Law* clearly stipulates that for information push and business marketing to individuals via automated decision-making, options not specific to personal characteristics or convenient refusal methods should be provided, that processing sensitive personal information such as biometrics, medical care, financial accounts and whereabouts must obtain individual consent, and that applications that illegally process personal information shall be ordered to suspend or terminate the provision of services.

7.3.3 The Special Rectification of Personal Information Protection on Apps Achieves Positive Results

Presently, all types of Apps develop fast and become a key field of personal information protection, with the number of available Apps and the scale of users continuously expanding. The CPC Central Committee and the State Council attach great importance to the protection and governance of personal information on Apps. The Cyberspace Administration of China, the Ministry of Industry and Information Technology, the Ministry of Public Security and the State Administration for Market Regulation continue to deepen the governance of illegal collection and use of personal information on Apps.

In March 2021, the Cyberspace Administration of China, the Ministry of Industry and Information Technology, the Ministry of Public Security and the State Administration for Market Regulation jointly issued the *Provisions on the Scope of Necessary Personal Information for Common Types of Mobile Internet Applications*. As it clearly stipulates, mobile Internet app operators shall not forbid user's to use basic function services of the app on the excuse that users do not agree to collect unnecessary personal information. This effectively tackles the common problems of unauthorized or excessive collection of data as well as compulsory authorization and excessive claims for authorization, which arouses wide social response. In April 2021, the Cyberspace Administration of China, together with the Ministry of Industry and Information Technology, studied and formulated the *Interim Provisions on the Protection and Management of Personal Information on Mobile Internet Applications*, and canvassed public opinion. Meanwhile, the above-mentioned four departments further strengthened the coordination and collaboration in reporting violations and certification management. After May 2021, the Office of the Central Cyberspace Affairs Commission conducted a special inspection and evaluation of 1,184 Apps of 14 common types, such as input method, map navigation, security management, online lending and browser, which were downloaded by users in large quantities, and announced 351 Apps with serious violations of laws and regulations, ordering them to rectify within the time limit. Apps that were not rectified within the specified time limit would be removed, which effectively curbed illegal activities and achieved satisfactory results and positive social response.

7.3.4 The Rules on Judicial Protection of Personal Information Are Clarified

In terms of the civil protection of citizens' personal information, judicial authorities deeply explore the protection of the rights and interests of personal information, and provide a civil remedy for the leakage of personal information. For example, in the case of Ling's suing against Beijing Microlive Vision Technology Co., Ltd., in July 2020, Beijing Internet Court announced the judgment of the first instance, ruling that TikTok App processed Ling's personal information without Ling's consent, which constituted an infringement. In March 2020, Beijing Third Intermediate People's Court made a judgment on the second instance in the case of the tort liability dispute between Deng and SF Express, providing the civil remedy for personal information leakage.

The protection of biometric information comes to a crucial stage. Face recognition has become a hot topic in recent years. In reality, the malpractices of obtaining face recognition data without notice and using face recognition in a mandatory way often occur. The first case of face recognition in China is a service-contract dispute that arises from the collection and use of biometric information to verify identity by Hangzhou Wildlife World Co., Ltd. In the second instance, the judge added a judgment to delete fingerprint-identification information. As sensitive personal information, biometric information deeply reflects the physiological and behavioral characteristics of natural persons and has strong personality attributes. Once widely leaked or illegally used, biometric information probably results in discrimination against individuals or illegal harm to personal and property security. Therefore, biometric information should be cautiously processed and strictly protected. In terms of systematic construction, in April 2021, the National Information Security Standardization Technical Committee released the *Security Requirements for Information Security Technology and Face Recognition Data* and solicited public opinion. It mainly solves the problems of face-data indiscriminate collection, leakage or loss, as well as excessive storage and use. In July 2021, the 1841st plenary session of the Supreme People's Court Judicial Committee deliberated and adopted the *Provisions of the Supreme People's Court on Several Issues Concerning the Application of Law in the Trial of Civil Cases in Relation to the Use of Face Recognition Technology to Process Personal Information.* The *Provisions* clarifies the nature and responsibility of the abuse of face recognition technology to process face information from the perspective of personality rights and tort liability, and respond to issues of general concern to the people from the perspective of property services, the effectiveness of standard terms, and liability for breach of contract.

7.4 The Law-Based Cyberspace Governance is Strengthened and Cyber Ecological Environment is Purified

Cyberspace means the common spiritual home to hundreds of millions of people. A healthy cyber ecology is the prerequisite or guarantees for the public to participate in cyber activities in a regular and orderly way. It is the common responsibility of governments, enterprises and users to improve the governance of cyberspace ecology. The *Regulations on the Governance of Network Information Content and Ecology* has been officially implemented, which further improves the comprehensive cyberspace governance system and purifies the cyber ecological environment. Besides, the law-enforcement in cyberspace fully implements the idea of "comprehensive governance" and steadily promotes joint law-enforcement.

7.4.1 The Legal System of Minors Internet Protection is Improved

In October 2020, the 22nd Session of the Standing Committee of the 13th National People's Congress deliberated and adopted the newly-revised the *Minors Protection Law of the People's Republic of China*, which would come into force on June 1, 2021. As the basic legislation in the field of minors protection in China, grounded in the status quo of minors' Internet protection, the *Minors Protection Law* adds a chapter of "Internet Protection" and forms a solid legal guarantee for minors' Internet protection in China. As required, Internet service providers in the areas of online games, online livestreaming, online audio & video and online social media should possess functions like time management, authority management and consumption management when minors use their services. Involving governments, schools, families, network products and service providers, this chapter stipulates the content of network literacy education, network information content management, personal information protection, network addiction prevention and network bullying prevention, aiming to achieve the online and offline protection of the minors in an all-round way. The addition of the chapter "Internet Protection" marks a milestone in better protecting the legitimate rights and interests of minors in the Internet era, demonstrates the legislative goal of Internet protection of children, actualizes the application of the *Cybersecurity Law* into the protection of the minors, and highlights the optimization and systematization of legal standards in China's cyber society.

In March 2021, Yuhang District Procuratorate (Hangzhou, Zhejiang) filed a civil-public-interest lawsuit against a well-known short video company in China, which infringed on children's personal information. The case was closed after Hangzhou Internet Court issued a mediation letter. The case signals the first civil-public-interest lawsuit filed by the procuratorial organ in "minors Internet protection" in China after the implementation of *The Civil Code* and the revision of the *Minors Protection Law*. The case focuses on the implementation of the provision in the *Provisions on the Internet Protection of Children's Personal Information* that "Internet operators who collect, use, transfer and disclose children's personal information should inform children's guardians in a significant and clear way, and should obtain the consent of children's guardians". This means both the exploration of procuratorial organs in the civil-public-interest litigation of minors' Internet protection and practice in the implementation of the minors-Internet-protection standards in judicial procedures.

7.4.2 The Information Content in Cyberspace is Regulated

In January 2021, the Cyberspace Administration of China introduced the *Measures for the Administration of Internet Information Services (Draft Revision for Comments)*, a basic regulation in the field of Internet management. The *Measures* explicitly prohibits paid deletion of posts, false transactions, account reselling, infringement of other's reputation, privacy and intellectual property rights, as well as the fabrication and dissemination of false information related to the COVID-19 pandemic. Whoever refuses to rectify or seriously violates regulations will be fined 100,000–1,000,000 *yuan*. The revised *Measures* serves as the regulatory framework for governing Internet information services under new circumstances.

In 2021, in the work report of the Supreme People's Procuratorate in the "Two Sessions", the procuratorial system actively participated in cyberspace governance. In 2020, procuratorial organs sued 142,000 subjects of cybercrimes, with an increase of 47.9% year on year against the background of the decline in the total number of criminal cases. As traditional crimes quickly spread to the Internet, procuratorial organs set up guidance groups on procuratorial cases, laid down 65 standards for prosecuting and accusing criminals, made full use of the system of professionals-aided case handling, and sped up law-based cyberspace governance. In line with the needs of handling cases, procuratorial organs offered procuratorial suggestions to relevant departments to jointly bolster comprehensive governance of cyberspace order, in terms of regulating Internet black & grey industrial chains, improving mobile

Internet's capacity in regulation and law-enforcement, and strengthening the Internet protection of the minors.

7.4.3 Online Livestreaming Marketing Information Industry is Regulated

In recent years, as a new Internet business model, online livestreaming marketing developed rapidly and played a positive role in promoting employment, expanding domestic demand, boosting the economy and alleviating poverty. Simultaneously, various problems arise, such as anomie in the words and deeds of online livestreaming marketing personnel, profits from minors as livestreaming hosts, inadequate performance of the platform's main responsibilities, false publicity & data fraud, frequent occurrence of counterfeit and shoddy goods, and difficulties in consumer rights protection and evidence collection. This necessitates the formulation of corresponding systems and regulations timely. In April 2021, the Cyberspace Administration of China and the other six departments jointly issued the *Measures for the Administration of Online Livestreaming Marketing (For Trial Implementation)*, which would come into force on May 25, 2021. The *Measures*, as an important administrative normative document for the implementation of the *Cyber Security Law of the People's Republic of China*, the *E-commerce Law of the People's Republic of China*, the *Advertising Law of the People's Republic of China*, the *Anti-Unfair Competition Law of the People's Republic of China* and the *Provisions on the Ecological Governance of Network Information Content*, possesses great practical significance in regulating the order of Internet market, protecting the legitimate rights and interests of people, promoting the healthy and orderly development of new business forms, and creating a clear cyberspace. The *Measures* clarifies and optimizes the respective rights and responsibilities of the participants such as online livestreaming marketing platforms, livestreaming room operators, livestreaming marketing personnel and service providers for livestreaming marketing personnel, and further enhances the responsibilities of relevant parties.

7.4.4 Internet Law-Popularization is Further Promoted

In recent years, the Cyberspace Administration of China and its relevant departments continuously strengthened the publicity and education of Internet law-popularization, took law-abiding and law-popularization as the basic work of law-based governance of cyberspace, realized the normalization of law-popularization, and guided the masses to reinforce their awareness of law in cyberspace. In September 2020, China launched the Cybersecurity Week, under the theme of "Cybersecurity serves the people and depends on the people". In many places, special law-popularization was conducted in relation to cybersecurity in the forms of lectures, forums and summits, and a series of themed day activities were staged for law-popularization personnel, such as school day, telecom day, law day, financial day, youth day and personal information protection day. This highlights the theme of law and improves public awareness of the law in cybersecurity. In March 2021, the Office of the Central Cyberspace Affairs Commission and the Ministry of Justice jointly convened a national teleconference on Internet law-popularization, reporting the implementation of the "7th Five-Year Plan" law-popularization in national cyberspace affair system and the work deployment of Internet law-popularization in 2021. As required, relevant departments should take Xi Jinping thought on law-based governance as fundamental guidance and action guide, and adhere to the working principles of advancing in a coordinated and orderly way, relying on the people & serving the people, making innovations on the basis of tradition and stressing actual effect. Besides, relevant departments should improve the working mechanism for Internet law-popularization, innovate the methods of Internet law-popularization and create a new situation for Internet law-popularization. In April 2021, the Office of the Central Cyberspace Affairs Commission held a national conference on law-based cyberspace governance in Chongqing, which researched and deployed law-based cyberspace governance in the present and future. The conference required relevant departments to draw the blueprint for Internet law-popularization, actively promote the law via cases, give priority to the publicity and education of law in cyberspace among the youths, and forge a good atmosphere of respecting the law, observing the law and using the law in cyberspace among the society.

7.4.5 The Regulation of Internet Public Welfare is Augmented

In charity and Internet regulations, it is an important task to combat illegal activities and strengthen the regulatory efforts in the field of Internet public welfare. In 2021, China stepped up efforts to crack down on illegal social organizations. In March 2021, the Office of the Leading Group for Combating and Punishing Illegal Social Organizations of the Ministry of Civil Affairs released the *Suggestions on Preventing Illegal Social Organizations from Cheating and Gaining Money in the Name of Public Welfare*. The document unmasks that it is illegal for illegal social organizations to conduct public fund-raising activities by publishing fund-raising information through radio, television, newspapers, the Internet and other media. Besides, in order to widely guide and mobilize the Internet industry to participate in the combat and rectification against illegal social organizations, the Ministry of Civil Affairs, working with the Office of the Central Cyberspace Affairs Commission and the Ministry of Industry and Information Technology, organized a symposium on the work of the combat and rectification against illegal social organizations in collaboration with Internet platforms, requiring all Internet platforms to improve their systems and specifications, strengthen information verification, provide no services to illegal social organizations, give full play to the advantage of information-data retrieval technology, and assist relevant departments in the rectification of illegal social organizations. 28 Internet platform enterprises including Xinhuanet, Social Assistance of China, Tencent, Sina and JD attended the symposium.[1] Simultaneously, social organizations in the field of Internet public welfare and Internet public donation information platform continue to reinforce industrial self-discipline and self-supervision. The China Internet Development Foundation and other major pivotal social organizations play a leading role in advocating Internet public donation information platforms to sign self-discipline commitments and promoting the standardized and orderly development of the Internet public welfare industry.

7.5 The Rules on Digital Governance Mature

The order in cyber society plays an important part in the overall social order, and an orderly cyber society guarantees the operation of the social system. In the governance of niche fields like digital economy, platform competition, online transaction, online livestreaming and smart judiciary, relevant legal work has been carried out systematically, with remarkable progress made in the fields of legislation, law- enforcement and judicature.

7.5.1 The Healthy Development of the Digital Economy is Regulated

In recent years, the platform economy in China thrived, and new business forms and models continuously emerged, which played an important role in promoting high-quality economic development and meeting the growing needs of people for a better life. Simultaneously, with advantages in data, technology and capital, the Internet economy displays a trend of high-degree concentration. Market resources intensively flow to giant Internet platforms and cause an increasingly prominent problem of Internet platform monopolies, which brings hidden risks and dangers to maintaining fair competition among enterprises and the protection of consumers' legitimate rights and interests. The meeting of the CPC Central Committee Political Bureau and the Central Economic Work Conference clearly require strengthening anti-monopoly and preventing the disorderly expansion of capital, which receives wide attention and extensive support among the public. Anti-monopoly and anti-unfair competition

[1] Gongyi Shibao: "To Crack down on Illegal Social Organizations, the Ministry of Civil Affairs Takes Various Measures, in which Industrial Associations, Internet Enterprises and Iconic Conference Venues Participate". April 1, 2021, https://mp.weixin.qq.com/s/E5YWt8pV-DxXKfF3BimgDQ.

become the internal requirements for improving the socialist market economic system and promoting high-quality development.

In 2020, in the judicial work, relevant departments vigorously fostered the healthy development of the digital economy, clarified the rules for the development of the digital economy and technological innovation with law-based fair judgment, and guided the healthy and orderly development of new technologies and new business forms & models on the track of law. They also reasonably determine the boundaries of the platform's responsibilities and behaviors, and enormously boost the development of the platform economy and sharing economy in accordance with law. They hear the case of pay-per-view in advance on video websites, standardize the innovation on the business model, and protect the legitimate rights and interests of users. Besides, they strengthen the protection of the legitimate rights and interests of employees in new business forms like parcel and food couriers and online car-hailing drivers, hear cases of mobile phone software infringing user's personal information and face recognition disputes, enhance personal information protection, and protect data security.

In 2021, national legislation and law-enforcement agencies strengthened the governance of Internet platforms, highlighted the responsibilities of Internet platforms, and cracked down on the monopoly of Internet platforms. In February 2021, the Antitrust Commission of the State Council officially released the *Anti-Monopoly Guide in the Platform Economy Field*. As the first systematic anti-monopoly guide that specifically targets the Internet platform economy in China, it provides Chinese wisdom for anti-monopoly law enforcement in the world Internet platform economy. In response to the unfair competition on the Internet platform, in April 2021, the State Administration for Market Regulation imposed administrative penalties on Alibaba Group for its "one out of two" monopoly. As investigated, Alibaba Group enjoys a monopolistic and dominant position in the online retail platform service market in China. In line with the provisions of Article 47 and Article 49 of the *Anti-Monopoly Law of the People's Republic of China*, and taking into account the nature, degree, duration and other factors of Alibaba Group's illegal behavior, the State Administration for Market Regulation made administrative punishment on Alibaba Group according to the law, ordered it to stop its illegal behavior, and imposed a fine of 4% of its domestic sales (455.712 billion *yuan*) in 2019, i.e., a total of 18.228 billion *yuan*.

7.5.2 The Regulatory System of Online Transaction is Optimized

In the field of e-commerce regulation, the conditions for basic legislation are preliminarily complete, and key legislation is accelerated. In order to fully implement the *E-commerce Law of the People's Republic of China* and improve the regulatory system of online transactions, the State Administration for Market Regulation issued the *Measures for the Supervision and Management of Online Transactions* in March

2021, which would take effect on May 1, 2021. The *Measures* clearly stipulates operator's positioning in current social e-commerce, live commerce and other online trading activities, and strongly regulates new-type unfair competition behaviors such as fictitious transaction, misleading display & evaluation and fictitious traffic data, and prohibits all kinds of online consumption torts. As an important regulation for departments to implement the *E-commerce Law*, the *Measures* possesses great practical significance for improving the regulatory system of online transactions, continuously purifying online trading space, maintaining the online trading order of fair competition, and creating a safe and secure online consumption environment. To meet the regulatory needs of the online tourism industry, in August 2020, the Ministry of Culture and Tourism released the *Interim Provisions on the Management of Online Tourism Business Services*, which would come into force on October 1, 2020. The *Provisions* specifies the platform responsibilities of online tourism enterprises. In particular, it includes the long-standing problems in the industry that infringe on the legitimate rights and interests of consumers and disrupt market competition order into the scope of regulation, which has a far-reaching impact on the high-quality development of the tourism industry.

In July 2021, Keqiao District People's Court (Shaoxing, Zhejiang) heard the case of Ms. Hu suing Shanghai Ctrip Commerce Co., Ltd. for the infringement on her rights, which marked the first case of big data discriminatory pricing. After the hearing, the court held that Ctrip App, as an intermediary platform, had an obligation to truthfully report the actual value of the subject matter; yet, Ctrip App reneged on a deal. Ctrip promised the plaintiff that diamond-level VIPs enjoyed preferential prices. However, Ctrip did not take measures on price supervision and offered the plaintiff a false price with a premium of 100%, breaking its promise. Moreover, Ctrip informed the plaintiff of the reason why it could not return all the price differences when handling the plaintiff's complaint. After investigation, the reason was also at variance with the facts, with an industrial fraud. Therefore, the court ruled that the defendant indulged in false advertising and pricing fraud, supporting the plaintiff to "refund one and claim for three". The judgment suggests that the online trading market will be more stringently regulated at the judicial level, so as to better regulate operator's behavior and comprehensively protect the legitimate rights and interests of consumers.

7.5.3 Relevant Cases Provide Examples for the Clarification of Platform Data Rights and Interests

While determining the legal boundary between the collection and use of personal information and the use of other people's data, the judicial decision on the dispute over the ownership of corporate data provides an example of the protection of Internet personal information and data. In March 2021, Beijing Intellectual Property Court announced the judgment of the second instance on the case of the unfair competition between Beijing Weimeng Chuangke Network Technology Co., Ltd. and Hunan Eefung Software Co., Ltd. The court viewed in the judgment that the rights and interests claimed by the Internet platform, in addition to these of the data per se, included the cost control generated by the platform operation and the maintenance of data security, as well as the operating benefits obtained from the derivative utilization or development of the data based on the rights and interests of Internet platform. This affirms the efforts taken by Internet enterprises in data collection, development and classification at the judicial level, sets a model for the trial of similar cases, and plays a guiding role in balancing the relationship between the development of the digital economy and the protection of personal information and data.

7.5.4 *Smart Judiciary Keeps Pace with the Times*

In recent years, judicial departments in China voluntarily used cutting-edge information technologies like the Internet, big data, AI and blockchain to introduce products and services in smart courts and smart prosecutors and promote judicial modernization with judicial intelligence. For example, procuratorial organs adopt integrated business application systems, and Internet courts hear Internet platform cases, which arouse wide attention. The court system adheres to the integration of judicial reform and information construction. In 2020, the "Scales of Justice Project" passed the completion acceptance, marking a new step in the construction of smart courts. In the COVID-19 pandemic, smart courts played a significant role. The wide use of non-contact litigation services enabled litigants to participate in litigation without leaving their homes and achieved the goal that "trial and execution continued uninterrupted, so did fairness and justice".

Beijing Internet Court, Hangzhou Internet Court and Guangzhou Internet Court constantly make reforms and exploration in improving the litigation rules, innovating technological applications and promoting the law-based governance of cyberspace. They optimize the rules for online trials, electronic evidence and asynchronous trials, and ensure that online litigation is in accordance with the law. Averagely, an online trial takes 36 min and saves two-thirds of the time *vis-à-vis* offline trial; a case with an online trial takes 60 days, one-fourth shorter than that with the offline trial. Additionally, Internet courts make positive progress in innovating the trial of cases, quickening the construction of court informatization and building a diversified dispute resolution mechanism. It is a major duty of Internet courts to establish rules by judgment and promote governance by rules. For example, Hangzhou Internet Court has innovated judicial trial mechanisms, reshaped the trial rules of cases, and explored and formed a set of procedural rules that center on the operating procedures of online litigation platforms and the standards of online video trials. In 2020, Hangzhou Internet Court heard the case of traffic hijacking on mobile apps, punished the fraud of network traffic, and maintained the market order of fair competition. It also opened the judicial blockchain platform, supported online copyright owners to upload works and save evidence, and prevented & punished online plagiarism. By building a new world-leading Internet judicial model with Chinese characteristics, it contributes to a Chinese model for the development of global Internet law.

7.5.5 Internet Public Welfare Standard System is Constructed

The construction and improvement of the Internet public welfare standard system help to solve the problems such as the lack of norms for the behaviors of Internet public welfare subjects and the barriers to platform communication, and accelerates the formation of healthy Internet welfare ecology.

In recent years, various industrial and association standards in the field of public welfare (Internet public welfare) have been introduced and implemented. After the publication of four association standards, i.e., the *Code for the Management of Charity Organization Archives*, the *Code for the Management of Charity Organization Projects*, the *Guidelines for the Evaluation of the Three As and Three Forces of Public Welfare Projects* and the *Evaluation Method for Building Charity Communities*, in September 2020, China Charity Alliance released three association standards, i.e., the *Guidelines for the Information Disclosure of Charity Organizations*, the *Guidelines for the Press Release of Charity Organizations* and the *Operating Guidelines for the Selection of Venture Capital of Public Welfare*. Among them, the *Guidelines for the Information Disclosure of Charity Organizations* and the *Guidelines for the Press Release of Charity Organizations* involve the standards in the field of Internet public welfare. The *Guidelines for the Information Disclosure of Charity Organizations* is formulated to propel charitable organizations to better implement the *Measures for Information Disclosure of Charitable Organizations* issued by the Ministry of Civil Affairs, aiming to ensure the public right to know and supervise and enhance the social credibility of philanthropy.[2] The *Guidelines for the Information Disclosure of Charity Organizations* clarifies the detailed rules and guidelines for the publication of basic information, annual work reports, financial & accounting reports, charitable projects and major events of charitable organizations on the national charitable information platform built by the Ministry of Civil Affairs or Internet platforms independently-built by charitable organizations, as well as the publication of online fund-raising information of public organizations. The *Guidelines for the Press Release of Charity Organizations* proposes the principles, methods, organizational forms, after-the-event evaluation and precautions of charity organization's news release, as well as the necessary qualities of their press spokesmen and strategies for media communication, and clarifies a variety of news release channels and platform types, including the websites of the charity organization, the accounts of various new media and other Internet platforms. Industrial-supporting social organizations like the China Internet Development Foundation work together with relevant think tanks in the field of Internet public welfare, strengthen the construction of industrial

---

[2] Data Source: "China Charity Alliance Released Association Standards for Charitable Organizations". September 19, 2020, http://www.charityalliance.org.cn/News/14039.jhtml.

standards, and explore the establishment and improvement of the standards for the Internet public welfare industry from a non-governmental perspective. In August 2021, the China Internet Development Foundation and Institute for Philanthropy Development of Shanghai Jiao Tong University jointly released the *Research Report on China's Internet Public Welfare Standards*, which advocated that all actors in the field of Internet public welfare should participate in the formulation and improvement of standards, and contribute to the standardized development of China's Internet public welfare.

Chapter 8
Exchanges and Cooperation in International Cyberspace Governance

8.1 Outline

In the past year, COVID-19 occurred repeatedly. As the most serious infectious disease in a century, the pandemic raged globally. It was the primary task of international community to conquer the pandemic and reenergize economic growth as soon as possible. During the COVID-19 pandemic, new models such as telemedicine, online education, sharing platform and office automation were widely used. The Internet played an important role in boosting economic recovery, ensuring social operation and promoting international cooperation in combating the pandemic. Digital technologies were deeply integrated with the innovative application in relation to the operation & management in various fields of economy and society, in which cyberspace mingled and interacted with the physical world. In the face of the threats and challenges from unilateralism, protectionism and hegemonism to international order in the digital age, the international community placed greater expectations on major countries who should undertake more responsibilities in cyberspace. China achieved complete victory in the fight against poverty, implemented people-oriented development, comprehensively promoted the digital transformation and upgrading of the economy and society, accelerated the construction of a community with a shared future in cyberspace, and bore greater hope, with global attention and trust.

In November 2020, in his congratulatory letter to the World Internet Conference · Internet Development Forum, Chinese President Xi Jinping stressed that: "China stands ready to work with the international community to seize the historical opportunity of the information revolution, cultivate new momentum for innovation and development, create a new landscape of digital cooperation, establish a new pattern of cybersecurity, build a community with a shared future in cyberspace, and join hands to create a better future for mankind." More countries attach importance to seizing digital opportunities and seeking cooperation and development.

2021 meant the first year for China to implement the *Outline of the 14th Five-Year Plan for National Economic and Social Development and the Long-Range Objectives Through the Year 2035*, the 50th anniversary of China's restoration of its legal seat in

Chinese Academy of Cyberspace Studies, *China Internet Development Report 2021*,
https://doi.org/10.1007/978-981-19-9311-4_8

the United Nations, and the 20th anniversary of China's accession to the World Trade Organization, as well as an important year to create new landscape in international exchanges and cooperation in cyberspace. China advocates building a community with a shared future in cyberspace, actively advancing the process of international exchanges and cooperation in cyberspace, taking the initiative in seizing the opportunity to accelerate the formation of digital society, strengthening the planning of forward-looking issues in digital society governance, and constantly expanding new areas of international exchanges and cooperation. China supports the United Nations to play its role as the main channel, makes full use of international platforms to discuss relevant international rules, and continuously contributes Chinese wisdom to building peaceful, secure, open, cooperative and orderly cyberspace.

8.2 China's Participation in International Cyberspace Governance: The General Trend

Currently, China faces opportunities and challenges in all aspects of its participation in international cyberspace governance. With the normalization of the prevention & control of the COVID-19 pandemic, the demand for International cyberspace governance shoots up. The changes in international relations continually affect the ecology of international cyberspace governance. New technologies and new applications catalyze the combination and upgrading of cyberspace risks.

8.2.1 The Normalization of COVID-19 Prevention and Control Sharply Increases the Demand for International Cyberspace Governance

The COVID-19 pandemic poses unprecedented challenges to the global supply chain and industrial chain, cross-border information sharing and international technological cooperation, and exposes serious problems of emergency mechanisms and social governance systems in some countries. Simultaneously, the spread of the pandemic enormously accelerates the deep integration of physical space and cyberspace, reflects the low-level emergency response capacity of Internet governance, and further enlarges the demand for International cyberspace governance. The pandemic highlights the importance of international cooperation in developing the digital economy and ensuring global Internet connectivity. International cooperation plays a positive role in addressing cyberspace-governance issues and global crises.

At present, owing to the reality of inadequate supply for governance, international cyberspace governance shows a trend of supply–demand imbalance. In terms of public health, the digital divide has a more negative impact. Vulnerable groups and marginalized groups like the poor, the young, the elderly and women lack digital services and digital skills, depriving them of the opportunities to receive social relief and prevent viral infections and disasters. In terms of cybersecurity, global risks and hidden dangers continue to rise, and ransom-virus attacks, identity theft and other criminal activities occur more frequently. Even worse, hackers launch cyberattacks against medical, health and public service departments to steal vaccine information. In terms of social governance, on the one hand, the normalization of the pandemic stimulates the surge in digital applications, and exposes social problems such as the lack of regulation, the disconnect between rights and responsibilities, and the poor protection of data privacy. On the other hand, digital media platforms in some countries neglect their social responsibility, government regulation & public supervision mechanism need to be improved, and false information or hate speech aggravates the public opinion environment during the crisis of public health security. In terms of digital cooperation, the lack of international rules in cyberspace and the abuse of security technologies exacerbate the global trust deficit and the uncertainty of international order in cyberspace, and reshaping global digital trust and cyberspace order has become an important issue of common concern to the international community.

8.2.2 The Changes in International Relations Continuously Affect the International Governance Ecology of Cyberspace

Now, various governments play a prominent role in cyberspace governance, and the changes in international relations have a significant impact on the ecology of international cyberspace governance. On the one hand, the game between major countries intensifies, with substantial divergence among countries in the fields of national code of conduct, cybersecurity, digital rule and emerging technological standard in cyberspace. Cyberspace issues are intentionally securitized and politicized, which arouses wide concern among enterprises, research institutions and Internet users in various countries. On the other hand, the COVID-19 pandemic and extreme climate change hasten global economic recession. With the importance of global digital cooperation increasingly recognized, the international community embraces and upholds the idea of a community with a shared future in cyberspace in a deeper way.

As the most advanced scientific and technological country, the United States always coerces its allies and developing countries into picking sides and contains China's innovation and development, so as to maintain its dominant position in the digital field and eliminate potential threats. In a way, the United States becomes an important factor that undermines the international order in cyberspace. China and Europe face many challenges in deepening cooperation in the field of digital technology, yet they complement each other in the digital economy and green economy and forge environments for in-depth cooperation. Hammered by the pandemic, most developing countries lay their hope in economic recovery and call for deepening international cooperation in the digital field, expecting major countries to assume more responsibilities and provide more global public products to inject vitality into global economic recovery.

8.2.3 New Technologies and Applications Breed New Risks in Cyberspace

The continuous iteration of new technologies and the popularization of their social applications trigger a new-round technological revolution and industrial transformation. Emerging technologies, such as intelligent manufacturing, energy conservation & emission reduction, digital agriculture, health, AI, life sciences and blockchain-based cryptocurrency, are intertwined and integrated, taking the global scientific & technological revolution to a new stage and causing new governance problems or challenges to the international community.

In November 2020, the World Economic Forum (WEF) released a report entitled *Cybersecurity, Emerging Technology and Systemic Risk*. As the report suggests, with the development of emerging technologies, the scale, speed and connectivity of cyberspace are undergoing significant changes, which bring a series of new systemic risks and challenges. In January 2021, the WEF released *Global Technology Governance Report 2021: Harnessing Fourth Industrial Revolution Technologies in a COVID-19 World*, which specifically discussed governance challenges and solutions in key application fields such as AI, blockchain, IoT, digital travel and UAV. Taking 5G, IoT and big data as examples, owing to the rapid popularization of the commercial use of 5G, broadband network speed robustly supports Internet of Everything and the Internet of Humans in IoT cases. The number of user and device connected to the Internet grows at an unprecedented rate, and the data expand exponentially. Internet data not only constitute core production factors in the era of the digital economy, but also concern personal privacy, social development and national security, arousing a series of balance-relation problems that involve data security protection as well as data flow and sharing. Technologies and applications are constantly upgraded and developed, so late-industrializing economies probably overtake Internet development. In case of major breakthroughs made in the field of cutting-edge technologies or revolutionary technologies, the existing landscape in cyberspace will tremendously change. Therefore, international competition for new technologies and applications heats up, and the combination of technological and political factors further intensifies systemic risks.

8.3 Hot Topics in International Cyberspace Governance and China's Practice

Presently, the wave of information revolution is sweeping the world, profoundly altering the economic, interest, and security patterns of the world. Hot topics in cyberspace in 2021 show new characteristics. Various governance entities actively explore models in Internet platform governance, data governance and technology governance. As a builder, protector and contributor of the cyberspace order, China, together with the international community, takes a proactive attitude to jointly build cyberspace for the benefit of all mankind, and constantly conveys to the international community the courage and responsibility of a responsible major country.

8.3.1 Internet Platform Governance

In the Internet industry, the landscape of monopoly or "winner-takes-all" takes shape based on the number of users and data. Several giant Internet corporations monopolize massive data and Internet resources, and various countries extensively call for platform governance. Major countries and regions like China, the United States and Europe have successively taken measures to rectify Internet platforms. There are two issues that arise from platform monopoly.

(1) The abuse of data. For example, corporations engaged in social media and search engines like Facebook and Google tap massive resources of user's data, which are used to push and coordinate industrial production resources in a targeted way. When users deliver personal information in exchange for free services from Internet corporations, a large amount of behavioral data generated are collected by Internet corporations. This not only induces the problem of privacy protection but also amplifies the risk of large-scale data leakage.

(2) The monopoly of the market. Several platforms pocket massive data and form strong exclusive market positions, which produce a de facto access barrier and quicken the monopoly of the Internet market. In the long term, this will sabotage the healthy development of the Internet industry. Therefore, various countries take corresponding measures to strengthen the governance of platform monopoly. For example, members of the U. S. House of Representatives put forward a series of bipartisan legislation and related proposals that aim to harness the influence of large technological corporations, including the *Ending Platform Monopolies Act*, which involves technological corporations such as Google, Apple, Facebook, Amazon and Microsoft. The United States urges the finance ministers of G7 countries to reach two global tax treaties, intending to reduce the operating space of IT giants. After the *Ending Platform Monopolies Act* comes into force, the tax-avoidance structure built by some large technological corporations will not be completely ineffective, yet the tax-burden cost will be significantly increased.

As the only major economy that achieves positive growth against the headwinds of the pandemic, China remains at the forefront of the Internet platform economy in the world, and antitrust regulation further intensifies. The disorderly expansion of Internet platforms causes serious consequences, such as the risks in systematic monopoly, data security and social governance. It becomes a social consensus to improve laws and regulations as well as regulatory mechanisms, prevent & curb excessive monopoly in the field of the Internet platform economy, guide operators in Internet platform economy to obey the laws in businesses, safeguard the legitimate rights and interests of consumers and social public interests, and promote the sustainable and healthy development of the online economy. In December 2020, the Central Economic Work Conference was held in Beijing. General Secretary Xi Jinping attended and delivered an important speech. The conference emphasized that China would support the innovation-driven development of platform enterprises to enhance their international competitiveness, bolster the common development of the public and non-public sectors of the economy, regulate the development platform enterprises according to law, and optimize digital rules. Relevant departments should improve the legal standards on the monopoly verification of Internet platform enterprises, the collection, use and management of data, and the protection of consumer rights and interests. Besides, relevant departments should strengthen regulation, raise regulatory capacity, and resolutely combat monopoly and unfair competition. To address unfair competition, information monopoly and other issues, regulatory authorities in China summon Ant Group for talks, which helps to guide and standardize the operation and development of financial business. In February 2021, the Antitrust Commission of the State Council issued the *Antitrust Guidelines for Platform Economy*. The *Guidelines* systematically regulates the economic monopoly of Internet platforms and signals the stage of legalization and standardization in the antitrust governance of Internet platforms. In April 2021, the People's Bank of China, the China Banking and Insurance Regulatory Commission, the China Securities Regulatory Commission, the State Administration of Foreign Exchange and other financial management departments jointly conducted regulatory talks of 13 Internet platform enterprises in financial business. Additionally, the State Administration for Market Regulation imposed administrative penalties on Alibaba Group for its "one out of two" monopoly according to law. The above-mentioned regulatory measures further enhance Chinese Internet enterprises' awareness of fair competition and anti-monopoly compliance, and effectively facilitate the innovative, orderly and healthy development of the platform economy.

8.3.2 Data Governance

As data play a greater strategic role in the information era, many countries step up efforts to safeguard domestic data security, accelerate the formulation of international rules on cross-border data flow, and explore the construction of a governance framework. On the one hand, various countries place emphasis on the protection of personal data. For example, the European Union released the *General Data Protection Regulation* (GDPR) and established the model of personal data protection. Albeit the different measures taken in various countries, they adopt the GDPR principles in their legislations.[1] The United States endeavors to promote APEC's CBPR (Cross-Border Privacy Rules) and enhance the compatibility of personal information protection mechanisms. On the other hand, various countries establish "thresholds" for data security via "mutual recognition". The EU's GDPR determines the white list countries for the free cross-border flow of data through the rule of "adequacy decision", and gives "adequacy decision" to countries and regions with adequate protection level, so as to shortlist the countries and regions that can transmit data freely with the European Union.

China regards data security as the core issue in national development and international cyberspace cooperation in the "14th Five-Year Plan" and the future, and constantly strives to explore the governance framework that gives consideration to both development and security. In September 2020, China put forward the *Global Data Security Initiative* at an international seminar held in Beijing under the theme of "Seizing Digital Opportunities and Seeking Cooperative Development". In March 2021, spurred by the *Global Data Security Initiative*, China and the League of Arab States jointly released the *Initiative on China-Arab Data Security Cooperation*, reaching a high-degree consensus in the field of digital governance.[2] In 2021, China successively issued the *Data Security Law of the People's Republic of China* and the *Personal Information Protection Law of the People's Republic of China*, which provided a legal guarantee for data security and personal privacy protection. China makes steadfast progress in the governance of personal information protection and continues to augment its capacity in regulation and governance, which are fully recognized by more Internet users. In order to promote the healthy and orderly development of relevant industries, various departments successively publish industrial standards and documents, such as the *Guidelines for Health and Medical Data Security* and the *Guidelines for the Construction of Data Security Standard System in the Telecommunications and Internet Industries*. In March 2021, the Cyberspace Administration of China and other departments jointly issued the *Provisions on the Scope of Necessary Personal Information for Common Types of Mobile Internet Applications* to regulate the collection of personal information by mobile Internet apps, conduct professional testing on the collection and use of personal information of some apps widely used by the public, and accept public reports, which were widely welcomed.

[1] Huang Zhixiong and Wei Xinyu. "The Game on Cross-Border Data Flow Rules Between the United States and Europe and China's Response: A Perspective of the Invalid Judgment of *The Privacy Shield Agreement*". *Journal of Tongji University (Social Sciences Edition)*, 2021, 32 (02): 31–43.

[2] Data Source: The official website of the Ministry of Foreign Affairs, http://russiaembassy.fmprc.gov.cn/web/wjb_673085/zzjg_673183/jks_674633/fywj_674643/t1865097.shtml

8.3.3 Technology Governance

The technological innovation and national governance reform supplement and promote each other. The applications of new technologies solve knotty problems in national governance and mean new challenges to national governance.

As Industry 4.0 deepens, emerging technologies like big data, AI and blockchain will have a far-reaching impact on the development trend of cyberspace in the future. Scientific & technological innovation kindles the qualitative leap of productivity and the transformation of the socio-economic and national governance systems. New technologies and applications provide new ideas and tools for solving various problems in technology governance. Besides, technology governance serves as a major measure for China to prevent & control the COVID-19 pandemic and resume work and production. China attaches great importance to the development and governance of new information technologies. In December 2020, the Ministry of Industry and Information Technology issued the *Industrial Internet Innovation and Development Action Plan (2021–2023)* in order to deeply implement the innovation-driven development strategy of industrial Internet and promote the integration & development of industrialization and informatization in a wider range, deeper degree and higher level. In April 2021, the 4th Digital China Summit published the *Digital China Development Report 2020.* As confirmed, China scored great success in the prevention and control of the COVID-19 pandemic, in which digital technologies like Internet big data played a pivotal role, and digital monitoring & analysis strongly supported the accurate prevention & control of the pandemic. The Ministry of Industry and Information Technology gave full play to communications big data to prop up the monitoring & analysis of the pandemic. The E-government Office of General Office of the State Council promoted universal recognition of health codes nationwide and realized one-code pass. In June 2021, the Ministry of Industry and Information Technology and the Office of the Central Cyberspace Affairs Commission jointly released the *Guiding Opinions on Accelerating and Promoting the Application of Blockchain Technology and Industrial Development*. As the first guiding document for the blockchain industry issued at the national level, the *Guiding Opinions* helps to promote and standardize the development of China's blockchain industry.

8.4 China Enthusiastically Carries Out International Exchanges and Cooperation

Over the past year, China continued to deepen international dialogue and exchanges in cyberspace, adhered to the idea of extensive consultation, joint contribution and shared benefits, tightened up bilateral, regional and international dialogue and cooperation, engaged in establishing widespread cooperative partnerships with various parties in the international community, and sped up down-to-earth cooperation in the fields of the digital economy, cybersecurity, data security, Internet basic resources and cyber culture, so as to expand exchange & cooperation platforms in cyberspace and work together for a new era of digital civilization.

8.4.1 Building the Platform for International Cyberspace Governance

Since 2014, China has hosted the World Internet Conference yearly at Wuzhen, Zhejiang Province. The World Internet Conference has achieved fruitful results in ideological exchange, technological exhibition, economic & trade cooperation and building consensus, and successfully established an international platform for the connectivity between China and the world as well as a Chinese platform for joint contribution and shared benefits of international Internet. In November 2020, the World Internet Conference · Internet Development Forum was staged, and Chinese President Xi Jinping sent a congratulatory letter to the forum, which won a warm response from the guests. Centering on the theme of "Digital Empowerment for the Future: Building a Community with a Shared Future in Cyberspace Together", the guests participated in the forum in an offline or online way. They deeply discussed and widely communicated on various topics, demonstrating the idea of a community with a shared future in cyberspace.

At the forum, the Organizing Committee of the World Internet Conference released the action initiative *Jointly Build a Community with a Shared Future in Cyberspace*, which provided a feasible, practical path for building a community with a shared future in cyberspace, and further expanded and deepened international exchanges and cooperation in cyberspace. The action initiative follows President Xi Jinping's "Four Principles" on promoting the reform of the global Internet governance system and "Five Points" on building a community with a shared future in cyberspace, and proposes 20 actions from four aspects, i.e., achieving shared development, ensuring common security, realizing joint governance and enjoying benefits together. The action initiative calls on governments, international organizations, Internet enterprises, technological communities, social organizations and individual citizens in various countries to adhere to the global governance vision of extensive consultation, joint contribution and shared benefits, so as to build cyberspace into a development community, a security community, a responsibility community and an interest community that would benefit all mankind. Concerning the COVID-19 pandemic in the world, the action initiative specifically proposes to use the information and communications technology to improve the level of global public services and boost experience sharing and cooperation in using digital technology to respond to public emergencies such as the pandemic and natural disasters. Additionally, think tanks like Wuhan University, China Institutes of Contemporary International Relations and Shanghai Academy of Social Sciences jointly published *Sovereignty in Cyberspace: Theory and Practice (Version 2.0)* at the sub-forum that advocates building a just and reasonable international order in cyberspace based on national sovereignty and jointly building a community with shared future in cyberspace. Hopefully, under the framework of the United Nations, various countries should strengthen communication, implement equal consultation, seek common ground while shelving differences, and take a cooperative approach. On the basis of safeguarding national cyberspace sovereignty, various countries should formulate universally-accepted international rules and national codes of conduct in cyberspace, forge extensive consensus, contribute wisdom and strength, and jointly build peaceful, secure, cooperative and orderly cyberspace.

In August 2021, the China-Africa Internet Development and Cooperation Forum was held in the form of video links, under the theme of "Seeking Common Development, Sharing Cybersecurity, and Building a Community with a Shared Future in Cyberspace", hosted by the Cyberspace Administration of China. Representatives from 14 African countries and the African Union Commission attended the forum, who carried out in-depth exchanges on issues such as sharing digital-technology dividends and jointly safeguarding cybersecurity. At the opening ceremony of the forum, China proposed the *Initiative on China-Africa Jointly Building a Community with a Shared Future in Cyberspace*, calling on Chinese and African governments, Internet enterprises, technological communities, social organizations and individual citizens to join hands to build a closer community with a shared future in cyberspace.

8.4.2 Partaking in the Governance Progress Under the U. N. Framework

In recent years, various forms of dialogues and discussions have been launched on the formulation of international rules in cyberspace, and the call for the U. N. framework has increased. China has always actively participated in the process of international cyberspace governance and supported the United Nations in playing a central role in safeguarding cybersecurity and promoting international cyberspace governance. China constructively participated in the U. N. Information Security Open-Ended Working Group (OEWG) and Group of Governmental Experts (GGE), made a contribution to finalizing reports on the "dual processes" in March and May 2021 respectively, and advocated various countries to develop a comprehensive, transparent, objective and fair supply-chain-security risk-evaluation mechanism, establish global unified rules and standards, and promote the openness, integrity, security and stability of global IT product supply chain.

In September 2020, China released the *Position Paper of the People's Republic of China on the 75th Anniversary of the United Nations*, calling on the international community to strengthen dialogue and cooperation on the basis of mutual respect, equality and mutual benefit, use cyberspace for economic and social development, international peace & stability and global well-being, oppose cyber war and the arms race and foster a peaceful, secure, open, cooperative and orderly cyberspace. As stated, the pressing task is to develop an international code of conduct for cyberspace that is acceptable to all. China opposes any country's overstretching of the concept of national security to restrict normal ICT development and cooperation.[3] The 75th Session of the United Nations General Assembly adopted *A/RES/75/240 Resolution*, reiterating the need to build a community with a shared future in cyberspace, authorizing the launch of new-round Information Security OEWG, and discussing and formulating relevant rules on data security and other issues of universal concern. In April 2021, the 75th United Nations General Assembly High-Level Thematic Debate on Digital Cooperation and Connectivity was staged at the United Nations Headquarters in New York, under the theme of "Mobilizing the Whole Society to Bridge the Digital Divide". The representative of China attended the debate and delivered a speech in the form of videos, introducing China's positive measures to improve the level of digital cooperation and bridge the digital divide, and demonstrating China's proactive attitude in promoting digital cooperation and participating in digital governance and international rule-making.

The World Summit on the Information Society (WSIS) is the largest annual summit in the field of ICT in the world hosted by the United Nations. For six consecutive years, the Internet Society of China held seminars on information equity services and inclusive building at the WSIS. In 2021, the seminar was organized online, inviting Secretary General of the International Telecommunication Union, UNESCO representatives in China and other domestic and foreign guests to attend. The seminar possesses positive significance for helping the disabled with an information approach, bridging the digital divide, improving the environment for information services, and facilitating the disabled, the elderly and other social members to integrate into the information society. The China Academy of Information and Communications Technology (CAICT) develops and builds big data platform for universal telecommunications services, which effectively ensures the implementation of the pilot work of universal telecommunications services through the innovative application of information technology, and attains remarkable progress in narrowing the urban–rural digital divide, promoting rural development and eliminating poverty. In 2020, CAICT won the champion prize of the WSIS.

The Chinese government, enterprises, industrial associations, technological communities and other relevant parties have deeply participated in the United Nations Internet Governance Forum (IGF) for many years, and conducted extensive exchanges and discussions with representatives from global political, business and academic circles as well as non-governmental organizations. In November 2020, at the 15th United Nations Internet Governance Forum, the International Cooperation Bureau of the Cyberspace Administration of China and the Chinese Academy of Cyberspace Studies (CACS) jointly hosted an open forum under the theme of "Cyberspace Trust Mechanism Building in the Context of International Public Health Emergencies", which touched upon the impact of the COVID-19 pandemic on cyberspace governance, the role of Internet technological innovation in the fight against the pandemic and the path or approach of building a trust mechanism in cyberspace. Experts and scholars from top think tanks in China, the United States,

[3] Data Source: http://www.gov.cn/xinwen/2020-09/11/content_5542461.htm.

Russia, the United Kingdom, Germany, South Korea and other countries, as well as the winners of the Internet Hall of Fame, made keynote speeches, and representatives from more than 60 countries and international organizations, including the United Nations, the European Union, the African Union and the Asia–Pacific Network Information Center, took part in the discussion. The Cyber Security Association of China, in collaboration with the China Internet Development Foundation, ChinaEU, the European Council and Association France Internet Citoyen, hosted a seminar. Experts and scholars in the field of global cybersecurity and information protection& governance had an in-depth discussion on the topic of "personal information and privacy protection in the process of global COVID-19 prevention & control". The China Federation of Internet Societies and the United Nations International Children's Emergency Fund jointly organized the seminar on "the research and development governance of online games from a comprehensive perspective of children's rights". Besides, the Communication University of China and the University of Melbourne co-staged the seminar on "the impact of mobile Internet on the environment in the 5G era".

In February 2021, the 46th Session of the United Nations Human Rights Council was held in Geneva. The representative of the China Federation of Internet Societies delivered a speech at the annual full day meeting of "Children's Rights and Sustainable Development Goals", and issued a written speech on promoting online poverty alleviation at the meeting of "Boosting Poverty Alleviation and Protecting Human Rights" to the participants. China demonstrated to the international community the achievements of Chinese social organizations in promoting poverty alleviation and children protection, and contributing a China model to global poverty alleviation and children online protection, and advocated jointly building a community with a shared future for mankind and creating a more sustainable world.

8.4.3 Actively Strengthening the Exchanges and Cooperation in the Digital Economy

As a proponent of the development of global digital economy, China voluntarily participates in multilateral cooperation in the digital economy and deepens exchanges and cooperation with other countries in the digital economy. China plays a positive role in promoting the digital economy under multilateral mechanisms such as G20, APEC and BRICS. President Xi Jinping's speech at the G20 Riyadh Summit profoundly expounded China's propositions on the development and governance of the digital economy. The Summit released the *Leaders' Declaration of G20 Riyadh Summit* and the *G20 Digital Ministers Declaration*, which called on various countries to create an open, fair, just and non-discriminatory development environment, widely promote inclusiveness, highlight people-centered development, strengthen mutual benefit, and bridge the digital divide. China's major initiatives and measures robustly augment multilateral cooperation in global digital economy and inject strong momentum for the international community to overcome the COVID-19 pandemic and return to the track of economic growth and sustainable development. In 2020, China submitted to the APEC Digital Economy Steering Group the *Initiative on Using Digital Technologies to Help Prevent and Control the COVID-19 Pandemic and Economic Recovery* and *Initiative on Optimizing the Digital Business Environment and Activating the Vitality of Market Players*, which were unanimously approved by APEC Digital Economy Steering Group. In November 2020, in his speech "Working Together for an Asia–Pacific Community with a Shared Future" at the 27th APEC Economic Leaders' Meeting, President Xi Jinping reiterated relevant ideas in the two initiatives, which foregrounded China's leading role in international cooperation in the digital economy. In June 2021, the Cyberspace Administration of China, together with the Ministry of Foreign Affairs and Guizhou Provincial People's Government, organized APEC Digital Poverty Alleviation Seminar. A total of nearly 200 attendees participated in the seminar in online or offline way. The representatives came from APEC Secretariat, APEC member economies such as Chile, Malaysia, Mexico, New Zealand, Papua New Guinea, Russia and Thailand, relevant international organizations and industrial associations such as the World Economic Forum, Global System for Mobile Communications Association and U. S.-China Business Council, embassies in China such as Russia and Peru, and relevant Chinese governmental institutions, think tanks and enterprises. The participants spoke highly of China's contribution to the cause of global poverty alleviation, and the seminar scored great success. China vigorously builds BRICS Innovation Base for New Industrial Revolution Partnership, and leads and enlarges BRICS cooperation in the digital economy, digital innovation, digital transformation and other fields.

China launches communication with Russia, Serbia, Greece, Croatia, Switzerland, Denmark and other countries and expands China-EU cooperation in the digital field. In September 2020, China-EU High-Level Dialogue on Digital Cooperation was staged. Focusing on strengthening cooperation in the digital field, the two sides earnestly implemented the important consensus reached by Chinese and EU leaders, and held pragmatic and constructive discussions on ICT standards, AI and other topics.

China and ASEAN jointly advance innovation and cooperation, seize the opportunity of digital transformation, and build a China-ASEAN digital economy partnership. In November 2020, the Cyberspace Administration of China, the National Development and Reform Commission, the Ministry of Industry and Information Technology and Guangxi Zhuang Autonomous Region People's Government co-organized the 4th China-ASEAN Information Harbor Forum (Nanning). Under the theme of "Digital Technology Connects ASEAN, Intelligence Creates the Future", the forum gathered government officials, well-known experts and scholars and business elites from China, Laos, Vietnam, Malaysia and other ASEAN members, who discussed the development of the digital economy as well as smart connectivity, data exchange and mutual benefit in cooperation, provided suggestions for the construction of China-ASEAN Information Harbor, and helped China and ASEAN members to jointly build the "Digital Silk Road".

In April 2021, at the 4th Digital China Summit, under the guidance of the National Development and Reform Commission and the Cyberspace Administration of China, the China Center for Urban Development, the Office of Fujian Provincial Digital Fujian Construction Leading Group and Fuzhou Municipal People's Government co-sponsored "Digital Silk Road & China-ASEAN Smart City International Cooperation Sub-Forum". The sub-forum focused on "innovation, cooperation and common development". Government officials, experts & scholars and corporate representatives from China and ASEAN members exchanged their ideas on such topics as promoting regional trade and investment and optimizing mechanisms, policies, platforms and applications, so as to jointly boost China-ASEAN technological collaboration and innovation, share the achievements of smart-city construction, and plan the cooperation and communication in the future.

In August 2021, the Cyberspace Administration of China, the National Development and Reform Commission and The people's Government of the Ningxia Hui Autonomous Region co-organized the 2021 Online Silk Road Conference in Yinchuan, Ningxia. As an important part of the 5th China-Arab States Expo, the conference took "Innovation-Driven Digital Economy" as the theme. In an online or offline way, guests from home and abroad deeply probed into the issues of cutting-edge digital technology, digital infrastructure construction and digital economy, in a bid to take new measures on the construction of the "Digital Silk Road" and seize new opportunities for digital development.

As the cornerstone of the development of the digital economy, central bank digital currencies (CBDC) develop fast. Various countries in the world successively accelerate the research and experiment of central bank digital currencies, and the rise of China's e-CNY attracts wide attention. In line with relevant working-group mechanisms under the jurisdiction of the Bank for International Settlements, Financial Stability Board and International Monetary Fund, the People's Bank of China actively tackles the CBDC-related issues, researches and formulates CBDC standards and principles, and deeply explores the feasible path and regulatory mechanism to address the challenges of global currency stability. Additionally, led by BISIH Hong Kong, the Digital Currency Institute of the People's Bank of China, the Hong Kong Monetary Authority, the Bank of Thailand and the Central Bank of the United Arab Emirates jointly launched mCBDC Bridge, which aimed to advance the application of CBDC in cross-border payments.

8.4.4 Continuously Initiating International Cooperation on Cybersecurity

China firmly maintains peace, security and stability in cyberspace. In recent years, China has made remarkable achievements in coordinating and tackling major cybersecurity incidents and conducting international cooperation on cybersecurity. China actively participates in ASEAN Regional Forum (ARF) Intersessional Meeting on Cybersecurity and the Trust Building Measures Research Group Meeting, and enhances the awareness of cybersecurity emergency response and information sharing & cooperation. In particular, in 2020, as China advocated, ARF Ministerial Meeting released the *Declaration on Jointly Promoting ICT Security in the Context of International Security*. In August 2020, the 8th China-Japan-South Korea Annual Meeting on Internet Emergencies was held online. China, Japan and South Korea shared their ideas on the trends and policies as well as technological development in the field of cybersecurity, offered cases of major cross-border incidents and suggestions on cooperation, and discussed the trend of cybersecurity incidents and countermeasures under the background of the COVID-19 pandemic. In October 2020, CNCERT, as an ASEAN dialogue partner, took part in cybersecurity emergency drills conducted by ASEAN members for 14 consecutive times, which effectively tested the emergency-response technologies and capabilities of computer-security emergency-response organizations in various countries in handling cybersecurity incidents, and enhanced the cooperation between ASEAN and partner countries in jointly ensuring cybersecurity. In December 2020, 2020 Video Conference of CNCERT International Partners for Emergency Response was staged online. More than 80 representatives from more than 30 organizations in nearly 20 countries and regions participated in the conference, and exchanged opinions on the theme of "Cooperation on Cybersecurity Emergency Response in the COVID-19 Pandemic". In January 2021, China and Indonesia signed the *Memorandum of Understanding on the Development of Cybersecurity Capacity Building and Technological Cooperation*. Both sides agreed to further strengthen cooperation in the field of cybersecurity. Additionally, Chinese organizations play an active role in the exchanges on cybersecurity launched by platforms such as APEC-TEL, APCERT and ITU-SG.

Cybercrime and cyberterrorism are common challenges to various countries in the world, which impel the international community to strengthen cooperation. China resolutely combats cyberterrorism, widely launches cyber anti-terrorism cooperation with Shanghai Cooperation Organization and ASEAN members, and actively participates in the formulation of international rules against cybercrime under the framework of the United Nations. In November 2020, the 20th Meeting of the Council of Heads of State of the Shanghai Cooperation Organization released the *Statement of the Council of Heads of State of SCO Member States on Safeguarding International Cooperation in the Field of Information Security* and the *Statement of the Council of Heads of State of SCO Member States on Combating the Use of the Internet and Other Channels to Spread Terrorism, Separatism and Extremism*, which systematically elaborated on the common positions on the cooperation in the fields of information security and cyber counter-terrorism. In April 2021, China, Russia and Vietnam jointly organized the first ASEAN Regional Forum Seminar on Combating Cybercrime, which deepened participants' understanding of strengthening international cooperation and addressing the threat of cybercrime. Chinese organizations continue to partake in the United Nations GGE Meeting on cybercrime, share with international community relevant practices on international cooperation and prevention of cybercrime, set forth propositions, and put forward constructive suggestions. In May 2021, the 75th Session of the United Nations General Assembly adopted the *Resolution on the Negotiation Arrangements of the United Nations Convention Against Cybercrime*, and decided to start the first negotiation in January 2022. As scheduled, China would play a constructive role in the negotiation of the convention, and work closely with various parties to formulate an authoritative and universal convention as soon as possible, so as to provide a pragmatic and effective approach for the international community to jointly face the challenges of cybercrime.

8.4.5 Strengthening Exchanges and Cooperation in the Field of Internet Basic Resources

Chinese technological communities energetically participate in the work of international organizations and platforms in the field of Internet basic resources, and continuously improve the international governance system of domain name resources. ccNSO (Country Code Names Supporting Organization) of ICANN (Internet Corporation for Assigned Names and Numbers) is a community organization via which ICANN manages and formulates ccTLD (country code top-level domain)-related policies, and promotes exchanges and cooperation between ccTLD administrators. In November 2020, a technological expert from China Internet Network Information Center (CNNIC) was elected as new director of ccNSO, playing an active role in ccNSO's relevant policy formulation, agenda setting and international exchanges and cooperation, and solidifying China's activeness and recognition in international technological community.

APTLD (Asia Pacific Top Level Domain Association) is a ccTLD organization in the Asia Pacific Region, which aims to promote the exchange of information on the development of Internet domain name application technologies among members and enhance their influence on the global Internet. Now APTLD has 50 official institutional members and 19 associate members. In February 2021, a representative of CNNIC were elected as new director of APTLD, which would make more contributions to further maintaining the relations in the international community and deepening regional exchanges and cooperation.

APNIC (Asia–Pacific Network Information Center) is one of the five regional Internet IP-address registration & management organizations in the world, responsible for the allocation and management of IP addresses in the Asia Pacific Region. All IP- address resources in China are authorized by APNIC. In March 2021, a representative of CNNIC was elected as executive member of APNIC, which demonstrated China's technological strength and management experience in the domain name field in the Asia Pacific Region, deepened the cooperation and exchanges with Internet registration authorities in the Asia Pacific Region, and provided access for China to further expand strategic cooperation with neighboring countries and BRI countries and regions.

Chinese universities and research institutions give full play to their advantages and actively participate in the activities of global Internet organizations. The research bases for international cyberspace governance of universities like Beijing University of Posts and Telecommunications, Zhejiang University and Tsinghua University voluntarily take part in the activities of international organizations like the Internet Engineering Task Force. In particular, six proposals from Tsinghua University were adopted by the IETF.

8.4.6 Promoting the Formulation of International Standards in the Field of Information Technology and Security

Chinese enterprises, research institutions and universities play a positive role in formulating international standards in the field of information technology and cybersecurity and promoting the development of emerging technologies. Chinese organizations continually strengthen cooperation with standard organizations like ISO (International Organization for Standardization) and IEC (International Electrotechnical Commission), as well as industrial organizations like AIM (Association for Automatic Identification and Mobility) and DIMC (Data Identifier Maintenance Committee). National Information Security Standardization Technical Committee (TC260) undertakes the technological work of the Cybersecurity Technical Sub-Committee of ISO and IEC Joint Technical Committee 1 (ISO/IEC JTC1/SC27). Since 2004, TC260 has participated in the SC27 international standardization work conference for 18 consecutive years and promoted the transformation of a number of cybersecurity standards independently developed by China into international standards. Several experts serve as the convener and liaison officer of SC27, and contribute Chinese wisdom to the formulation of international standards on cybersecurity.

In terms of the identity resolution system of industrial Internet, in June 2020, the China Academy of Information and Communications Technology (CAICT) was authorized by Association for Automatic Identification and Mobility as an international numbering organization (code "VAA") with the capacity to allocate global identification codes, together with large international organizations like International Article Numbering Association, Institute of Electrical and Electronics Engineers and Universal Postal Union. In January 2021, DIMC officially approved the exclusive international data identifier "15 N" for industrial Internet identification and granted CAICT management and maintenance authority. DIMC boasts the only management organization for the list of data identifiers in the world. The approval of "15 N" signals the authority to manage international data identifiers that DIMC gives to a Chinese organization for the first time. This not only evidences the recognition of China's industrial Internet identification resolution system by the international identification industry, but also possesses great significance for the international promotion of industrial Internet identification.

8.4.7 Promoting International Cooperation on Global Data Security

Data security risk means a common challenge to various countries in the world, which necessitates global cooperation and extensive consultation. In 2020, China put forward the *Global Data Security Initiative*, which highlighted the protection of critical information infrastructure and personal information, the storage and retrieval of corporate overseas data, the security of supply chain and other issues, and proposed constructive solutions to maintain global data security and cybersecurity. In March 2021, China and the League of Arab States held a video conference on China-Arab States data security cooperation. China and the League of Arab States jointly released the *Initiative on China-Arab Data Security Cooperation*. League of Arab States became the first regional organization in the world to jointly publish the initiative on data security with China. As a regional international organization with great influence, the League of Arab States actively supports and jointly releases the initiative with China, which fully proves that the initiative conforms to the trend of the times and the common aspirations of the international community. The initiative not only mirrors the common destiny of China and the League of Arab States, but also represents the high level of strategic cooperation between them. Simultaneously, it helps to produce fruitful results in the digital cooperation between the two sides, inject the wisdom and strength from developing countries into global digital governance, accelerate the healthy development of the digital economy, and forge a community with a shared future in cyberspace. China welcomes all parties to actively answer the call and participate in making global rules on digital governance.

8.4.8 Strengthening the Cooperation Among Internet Media as Well as Online Cultural Exchanges and Mutual Learning

Against the backdrop of the COVID-19 pandemic, Internet communication technologies affect the public opinion ecology and people's production & lifestyle with unprecedented speed, scope and depth, and bring wide space and new opportunities for cooperation among Internet media and Internet cultural exchanges. In December 2020, the "2020 China-Russia Internet Media Cloud Forum" was launched in Beijing and Moscow via video conference. The event was jointly guided by the Cyberspace Administration of China and the Ministry of Digital Development, Communications and Mass Media of the Russian Federation, co-sponsored by *China Daily* and the China Internet Development Foundation, and hosted by www.chinadaily.com.cn. Representatives of government departments, Internet media and Internet enterprises as well as experts and scholars from China and Russia participated in the forum and conducted in-depth discussions and exchanges under the theme of "The Role of Internet Media in the Pandemic". China vigorously boosts cultural exchanges on the "Digital Silk Road". In November 2020, the Ministry of Culture and Tourism of China and the Ministry of Culture and Youth of the United Arab Emirates hosted a series of activities themed "The Silk Road Cultural Journey: China Digital Culture Exhibition Week (The United Arab Emirates Station)", including China-United Arab Emirates Digital Culture Forum and China-United Arab Emirates Digital Cultural Innovation Exhibition. In June 2021, the first Silk Road Digital Museum was staged globally. The first group of more than 40 museums from 18 countries participate in the construction of the online platform that integrates multiple functions like digital collection, digital exhibition, digital knowledge and cloud curation, realizes the exchanges and sharing of cultural resources and serves the propaganda of the spirit of the Silk Road.

Building a community with a shared future in cyberspace outlines China's approach that tallies with contemporary trends, undertakes the responsibility of development, and calls for joint efforts to address risks and challenges and promote global cyberspace governance. The international community should maximize the consensus as soon as possible, seize the historical opportunity of the information revolution, cultivate new momentum for innovation and development, create a new landscape of digital cooperation, form a new pattern of cyberspace, and work together to forge a better future for all mankind.

Postscript

Currently, the tide of information revolution historically mingles with the strategic goal of national rejuvenation in the context of global change on a scale unseen in a century. In parallel with the cause of the Party and the country, the Two Centenary Goals continuously advance, and cyberspace work comes to a new stage. Under the background of the centenary of the Communist Party of China, cyberspace work continues the struggle over the past century and undertakes a glorious and arduous mission. With *China Internet Development Report 2021* (hereinafter referred to as "the *Report*"), we hope to deeply propagate and interpret Xi Jinping Thought on Socialism with Chinese Characteristics for a New Era, especially his important thought on building China into a strong cyberpower. Besides, the *Report* aims to comprehensively represent the status quo of China Internet development, systematically draw lessons from China Internet development, and reasonably envisage the future of China Internet development. By exploring the path of China Internet development, the *Report* will introduce China's experience and contribute Chinese wisdom to Internet development and governance in various countries in the world.

During the compilation of the *Report*, we received guidance and support from the Office of the Central Cyberspace Affairs Commission (hereinafter referred to as "OCCAC"). Leaders of OCCAC give us specific guidance on the *Report*. Relevant ministries & commissions, bureaus & units of OCCAC as well as the Office of the Provincial Cyberspace Affairs Commission provide us with strong support in necessary data and materials. The *Report* is launched by the Chinese Academy of Cyberspace Studies (CACS), and co-organized & co-compiled by the National Computer Network and Information Security Administrative Center, Beijing University of Posts and Telecommunications and the National Industrial Information Security Development Research Center. Experts and scholars, including Zhou Hongren, Gao Xinmin, Hu Zhijian, Zhao Guojun and Zhang Li, contribute their valuable suggestions. Main contributors are: Xia Xueping, Fang Xinxin, Li Yuxiao, Xuan Xingzhang, Li Yingxin, Zou Xiaoxiang, Liao Jin, Cheng Yifeng, Jiang Wei, Jiang Yang, Nan Ting, Wang Hailong, Li Xiaojiao, Xu Yanfei, Jiang Shuli, Tian Yuan, Chi Haiyan, Li Wei, Yuan Xin, Chen Jing, Xu Yu, Xiao Zheng, Zhang Qiyuan, Wu Wei, Zhao Gaohua, Shen Yu, Li Yangchun, Deng Jueshuang, Cai Yang, Jia Shuowei,

China Internet Development Report 2021,
https://doi.org/10.1007/978-981-19-9311-4

Liu Chaochao, Sun Luman, Lin Hao, Liu Zhuoyue, Yang Xuecheng, Sui Yue, Zhang Xuejun, Fu Wei, Wang Xiaoyuan, Sun Jie, Meng Qingguo, Wang Lida, Xie Xinzhou, Ding Li, Wang Xiaoqun, Jia Hui, Li Shaopeng, Shang Xixue, Han Haiting, and Li Yan.

The *Report*, albeit successfully published thanks to the strong support and considerable help from all sectors of society, is inadequate in terms of perspective and insight due to our limited research level, working experience, and tight deadline. Therefore, we ardently welcome valuable opinions and suggestions from governmental departments, international organizations, research institutes, Internet corporations, and social organizations across different sectors, home and abroad, to help us produce better reports in the future and contribute more wisdom and strength to the world Internet development.

Chinese Academy of Cyberspace Studies (CACS)
August 2021

Printed by Printforce, the Netherlands